Nonlinear Finite Element Analysis of Composite and Reinforced Concrete Beams

Woodhead Publishing Series in Civil and Structural Engineering

Nonlinear Finite Element Analysis of Composite and Reinforced Concrete Beams

Xiaoshan Lin
Y.X. Zhang
Prabin Pathak

ELSEVIER
WP
WOODHEAD PUBLISHING
An imprint of Elsevier

Woodhead Publishing is an imprint of Elsevier
The Officers' Mess Business Centre, Royston Road, Duxford, CB22 4QH, United Kingdom
50 Hampshire Street, 5th Floor, Cambridge, MA 02139, United States
The Boulevard, Langford Lane, Kidlington, OX5 1GB, United Kingdom

Notices

Knowledge and best practice in this field are constantly changing. As new research and experience broaden our understanding, changes in research methods, professional practices, or medical treatment may become necessary.

Practitioners and researchers must always rely on their own experience and knowledge in evaluating and using any information, methods, compounds, or experiments described herein. In using such information or methods they should be mindful of their own safety and the safety of others, including parties for whom they have a professional responsibility.

To the fullest extent of the law, neither the Publisher nor the authors, contributors, or editors, assume any liability for any injury and/or damage to persons or property as a matter of products liability, negligence or otherwise, or from any use or operation of any methods, products, instructions, or ideas contained in the material herein.

Library of Congress Cataloging-in-Publication Data
A catalog record for this book is available from the Library of Congress

British Library Cataloguing-in-Publication Data
A catalogue record for this book is available from the British Library

ISBN: 978-0-12-816899-8 (print)
ISBN: 978-0-12-816900-1 (online)

For information on all Woodhead publications
visit our website at https://www.elsevier.com/books-and-journals

Publisher: Matthew Deans
Acquisition Editor: Glyn Jones
Editorial Project Manager: Ana Claudia A. Garcia
Production Project Manager: Swapna Srinivasan
Cover Designer: Christian J. Bilbow

Typeset by SPi Global, India

Contents

Preface

Finite element method has been one of the most popular, powerful, and robust numerical modelling techniques for finding approximate solutions to practical engineering problems. In recent decades, with the knowledge advancement in science and engineering, composite materials, such as fibre-reinforced polymers (FRP), have been increasingly used for innovative design and retrofitting/repairing of structural components to enhance durability, sustainability, and resilience of infrastructures. Effective analysis and reliable prediction of the performance of composite structures especially those under extreme loading conditions, such as cyclic, fire, impact, and blast loads, is not an easy task, but it is essential and crucial for the structural design. Thus, there is a high demand of further development and advancement of numerical modelling methods and techniques.

This book provides comprehensive finite element analysis (FEA) procedures and modelling techniques for composite beams with a focus on the FRP-reinforced/strengthened concrete beams. In particular, the critical issues encountered in the structural analysis of composite FRP-reinforced concrete beams have been addressed by using the composite beam elements developed by the authors. Additionally, numerical techniques for establishing effective finite element model using the commercial FEA software ANSYS are introduced. The FEA source codes are also provided in this book. The readers would benefit from the ideas for the development of new elements, the modelling techniques applied in the analysis of composite beamlike structural components, and the source codes for FEA of composite beams. With the growing demand for reliable, accurate, and highly efficient numerical predictions, I believe this book forms a useful and valuable complement to many other excellent books already published in the field, and I wish the book can provide a valuable reference for researchers and engineers.

I have been very fortunate to have received excellent mentoring in the field of computational mechanics from world-leading high-profile scholars, particularly my PhD supervisor Professor Y.K. Cheung who has made pioneering contributions to the finite element method. My understanding and appreciation of the finite element method have been largely inspired by him, and I have been working on the finite element development and computer-aided numerical simulation of composite materials and structures in the past 20 years. I was also fortunate to have my first PhD student, a very bright and talented girl, Dr. Xiaoshan Lin, who is the first author of this book, to work with me on the FEA of composite beams. The majority part of this book is from her PhD research. The contribution from the third author Prabin Pathak, who was also my research student, is acknowledged.

Y.X. Zhang
Sydney, Australia
May 2019

Introduction

1

1.1 General introduction

1.1.1 Fibre-reinforced polymers in concrete structures

In civil and infrastructure engineering, the deterioration of structures can be caused by a variety of factors, such as corrosion due to marine environments, high chloride content in the air and the use of deicing salts on roads, alkali–silica reactions, poor initial design, poor construction and maintenance, and natural disasters such as earthquakes and hurricanes [1]. The added cost of repairing the deteriorated structures with replacement cost is usually more than twice the original cost of construction [2]. Hence, there is an increasing need for the development of advanced materials and techniques to improve the performance of civil engineering structures, prevent their premature deterioration, and prolong their service lives. In recent decades, composite materials have been successfully used in aerospace, marine, transportation, and civil construction industries. This stems from the fact that, through careful design, combinations of different materials adopt the advantages and eliminate the shortcomings of each individual component [3]. Fibre-reinforced polymer (FRP) has emerged as a promising composite material for enhancing the performance of reinforced concrete structures in civil constructions.

FRPs generally consist of synthetic or organic high strength fibres in a resin matrix. Those most commonly used for civil engineering structures are carbon fibre-reinforced polymer (CFRP), aramid fibre-reinforced polymer (AFRP), glass fibre-reinforced polymer (GFRP), and basalt fibre-reinforced polymer (BFRP) [2]. The general superiorities of FRPs to steel, which is the traditional reinforcing material for concrete structures, include

- excellent corrosion resistance and electromagnetic neutrality;
- high ratio of strength to mass density;
- excellent fatigue resistance, especially CFRP and AFRP;
- cost-effective fabrication;
- low axial thermal expansivity, especially CFRP.

With the excellent electrochemical corrosion resistance, FRPs are being widely used in the construction of concrete structures as a substitute for traditional steel reinforcing bars (rebars) where the corrosion usually takes place. In 1986, the world's first highway bridge prestressed with GFRP rebars was built in Germany. Since then, bridge constructions using FRP-reinforced concrete have been carried out throughout Europe, America, and Japan [4, 5]. In addition, FRP reinforcing bars have been applied in the constructions of industrial roof decks, highway barriers, chemical and wastewater treatment plants, sea walls, floating docks, and many other seafront structures since 1980s [4, 6]. Owing to their excellent dielectric properties, FRP

Nonlinear Finite Element Analysis of Composite and Reinforced Concrete Beams
https://doi.org/10.1016/B978-0-12-816899-8.00001-8

reinforcing bars have also been used in electrical substation reactor bases, magnetic resonance imaging facilities, airport runways, hospitals, and laboratories [4, 6].

However, FRP composites also have disadvantages, such as their relatively high cost, low elastic modulus, linear-elastic brittle behaviour, and poor fire resistance. To overcome the high cost of FRP materials, GFRPs, which are much cheaper than CFRPs, are commonly used in civil engineering constructions [7]. Cost savings also arise from various aspects, such as fabrication, maintenance, retrofit, and rehabilitation of structures, which actually offset their higher initial material cost. Whilst the low elastic modulus of FRP may cause large deflections in a structure at a relatively low stress level, and its linear-elastic brittle behaviour results in fragile rupture in composite FRP-reinforced/strengthened concrete members. Besides, at present, most applications of FRP composites are restricted to constructions in which the temperature effect is not a primary concern, which may be attributed to the fact that the mechanical properties of FRP deteriorate with increase in temperatures.

1.1.2 Critical issues in analyses of FRP-reinforced/strengthened concrete beams

Beam is one of the most basic structural elements as well as an indispensable part of most engineering structures. Nowadays, FRP-reinforced/strengthened concrete beams are increasingly being used in civil engineering constructions instead of traditional steel-reinforced/strengthened concrete beams. Therefore, a number of phenomena concerning the structural behaviour of conventional steel-reinforced concrete beams must be reconsidered, including the bond performance between reinforcing/strengthening material and concrete, as well as the effects of temperature and dynamic loads on the mechanical properties of materials and on the structural behaviours.

1.1.2.1 Bond behaviour

Bond between concrete and the reinforcing/strengthening material plays an important role in transferring stress from the former to the latter, and the bond behaviour is one of the most critical aspects of the performance of reinforced concrete structures. In FRP-reinforced concrete beams, the bond characteristics affect the anchorage of bars, the strengths of lap splices, and the required concrete cover [7]. Also, the performance of reinforced concrete beams depends not only on the material properties of concrete and reinforcing/strengthening material but also on the bond behaviour between the two components. The resisting mechanisms under bending, shear, and torsion are related to the development of an adequate bond at both serviceability and ultimate states [8]. In addition, bond length may affect the ductility of the concrete beams externally strengthened with FRP. Therefore, a proper design against debonding failure becomes a matter of concern for concrete beams reinforced/strengthened with FRP.

Due to the lack of well-established standards, a wide variety of FRP reinforcing bars, from those with simple and smooth surfaces to those with treated surfaces aiming to improve bond characteristics, have been commercialised. There are mainly two methods for enhancing the bond behaviour: deformation of the outer surface (ribbed, indented, and braided bars) and surface treatments (grain-covered bars) [8]. Besides, FRP bars can also be produced with varying compositions. Consequently, the bond between FRP reinforcing bars and the surrounding concrete is complicated and different from that of conventional steel reinforcements. Various factors may influence the bond characteristics of FRP reinforcements to concrete, including the shapes of the outer surfaces of FRP rebars, the coating with sand, the concrete compressive strength, the confinement pressure, the rebar diameter and positions in the cast specimen, the embedment length, the change of temperature, and the environmental conditions [8].

FRP laminates, which are typically made of continuous fibres, such as glass fibre, carbon fibre, aramid fibre, and basalt fibre, and bonded with a polymeric matrix like epoxy [9], are often used to strengthen concrete structures such as beams and panels. Although FRP composite plates can be used as an effective externally strengthening material, a sufficient bond between concrete and composite plate is essential to avoid the brittle debonding failure that has been identified to be one of the major failure modes [10, 11]. In general, the strength, stiffness, ductility, and failure mechanism of FRP-strengthened elements can be affected by types and area fractions of unidirectional composites, adhesive types, and bonding method.

In most analyses of FRP-reinforced/strengthened concrete beams, a perfect bond between reinforcing bars and the surrounding concrete (or between FRP laminate, adhesive, and concrete) has usually been assumed. This can provide a realistic simplification of a real bonding condition that requires sufficient surface preparation for the reinforcing bars/strengthening plates. However, with increase in load, cracking inevitably occurs, which results in a reduction in bond strength, and thus, bond–slip may take place, the effect of which on the structural behaviour of reinforced concrete beams cannot be ignored [12–14].

1.1.2.2 Effect of temperature

The resistance of reinforced concrete structures to fire is another extremely crucial issue that needs to be stressed. Since FRP materials are combustible and the mechanical and bond properties are susceptible to the elevated temperatures, there is a legitimate concern that FRP reinforcing/strengthening systems will perform poorly under fire conditions [15]. At present, FRP reinforcing bars and strengthening plates have mainly been used in structures where temperature effect is not a primary concern, such as bridges, whilst there is enormous economic potential for FRP materials to be applied in civilian constructions, parking garages, and industrial structures [16]. Thus, it is essential to investigate the structural behaviour of FRP-reinforced/strengthened concrete beams under fire conditions before implementing them in buildings and other fire-vulnerable structures.

The fire-resisting performance of conventional steel-reinforced concrete beams has been studied extensively in the past, and adequate fire endurance is usually ensured by providing minimum cross-sectional dimensions and sufficient concrete cover to the steel reinforcements to protect them from exceeding their prescribed critical temperature of 593°C [15]. However, this is not the case for concrete beams reinforced with FRP rebars, since no such critical temperature has been established for most currently available FRP rebars [15]. The behaviour of FRP-reinforced concrete beams under fire attack is different from that of conventional steel-reinforced concrete beams. When embedded in concrete, the burning of FRP reinforcements is avoided due to the lack of oxygen, whereas the resin will be softened. At the time when the temperature exceeds the matrix glass transition temperature, the mechanical properties of FRP deteriorate rapidly, resulting in reductions in its strength and stiffness, and consequently, the crack widths and the deflection of a FRP-reinforced concrete beam increase rapidly. When the temperature reaches a certain level, at which the fibres begin to degrade, collapse occurs. So far, as the available data are obtained from the experimental tests on FRP materials at various temperature levels and cover a wide range of fibre and matrix types, it is difficult to generalise the loss in mechanical properties of different types of FRP [17]. As a result, an appropriate temperature-dependent material model that can describe the deterioration of FRP mechanical properties at elevated temperatures becomes essential in structural analysis.

1.1.2.3 Effect of cyclic load

Reinforced concrete structures are very often subjected to cyclic loads in their service life, such as traffic and seismic loads. Under cyclic loads, the plastic deformation of a reinforced concrete element increases. The stresses in each constituent material and concrete–FRP interface are likely to increase with the increasing beam plastic deformation, which may promote brittle debonding and unexpected material failures [10]. Therefore, in addition to the strength of reinforced concrete components that have suffered material decay and damage under static load, it is necessary that they have sufficient durability to prevent failure from fatigue [18].

Fatigue life is defined as the number of cycles that lead to the failure of structural system [19]. The repeated loading would change the fundamental properties of a material, resulting in progressive damage propagation. For reinforced concrete structures, cyclic load typically causes softening behaviour of reinforcing steel and concrete [20]. Similarly, the mechanical properties of FRP are also affected by the cyclic load. Typical damage of FRP composites due to fatigue includes matrix cracking, debonding, and fibre fracture [21].

FRP strengthening plates normally consist of unidirectional fibres and matrix resin, and the fatigue response of FRP composites was reported to be dependent more on the matrix resin than the constitutive fibres [22], as the damage may propagate along the matrix between the unidirectional fibres. However, this is not the case for FRP-strengthened reinforced concrete components. When FRP is bonded on the surface of concrete, the failure of resin seldom occurs as the resin strength is usually greater than that of concrete in tension [19]. As the number of cycles increases, the fatigue damage is accumulated, causing the decrease in the friction between concrete and FRP plate and gradual bond failure, which indicates a progressive delamination of FRP.

The most commonly observed fatigue failure mode of a FRP-strengthened reinforced concrete beam is the rupture of the tensile reinforcing steel, followed by FRP strengthening plate [19]. This is attributed to the fact that, compared with

FRP plate, steel reinforcement is more susceptible to fatigue failure. Although the fatigue failure is primarily governed by steel reinforcement, FRP extends the fatigue life of the reinforced concrete beam by reducing stresses in the steel reinforcement. The delamination of FRP often occurs immediately after the rupture of steel reinforcement. Owing to the complex failure mechanisms and the various parameters that may affect the flexural fatigue life, it is still difficult to accurately predict the fatigue behaviour of reinforced concrete beams strengthened with FRP.

1.2 Scopes and structure

Nowadays, finite element method has become an important tool in the design and analysis of reinforced concrete structures. Commercial finite element software packages, such as ANSYS, ABAQUS, LS-DYNA, ADINA, and DIANA, have been widely used by researchers and engineers for various analysis purposes. Although the history of finite element method can be traced back to early 1940s [23] and it has become a mature numerical method over years' development, people still encounter difficulties when using finite element method in research and engineering practice especially for the analysis of complex structures under extreme loading conditions, including the selection of element types and material models, contact definitions, application of boundary conditions, and selection of nonlinear solutions. This book focuses on the finite element modelling of composite and reinforced concrete beams, aiming to provide some guidance to the readers in the numerical simulation of FRP-reinforced and FRP-strengthened concrete beams under various loading conditions.

More specifically, in this book, a family of simple shear-locking-free 1-D composite beam elements are introduced, which can be used to accurately and effectively model the structural behaviour of steel-/FRP-reinforced concrete beams under both ambient and fire conditions. The effect of bond–slip is taken into account in one of these composite beam elements. In addition, finite element analysis of reinforced concrete beams strengthened with FRP composite under cyclic loads is discussed. The commercial finite element software ANSYS is utilised in the numerical prediction. The modelling techniques for fatigue analysis, especially the simulation of delamination of FRP strengthening plate, are introduced. The outlines of this book are as follows.

Chapter 2 provides a general review of the finite element analysis of beam-like structural components. Basic beam theories, along with their merits and demerits, are discussed, followed by the important issues involved in the finite element analysis of beams and the modelling techniques. Nonlinear finite element analysis of beams and the solutions to the nonlinear finite element equations are also introduced.

In Chapter 3, a 1-D two-node composite beam element is introduced for the analyses of homogeneous isotropic and laminated composite beams. This element can be used to model both slender and deep beams without causing shear-locking problem. The basic formulations, displacement functions, and finite element equations for the composite beam element are discussed. At the end of this chapter, numerical examples are given to illustrate the finite element analysis using this composite beam element.

The composite beam element developed in this chapter forms the foundation for the analysis of FRP-reinforced concrete beams to be introduced in the following chapters.

In Chapter 4, nonlinear finite element analyses of steel-/FRP-reinforced concrete beams are discussed. A further developed composite beam element, including both geometric nonlinearity and material nonlinearity, is introduced and employed in the finite element simulation. The basic formulations for the element are firstly presented, and then, the material models of concrete, steel, and FRP are discussed, followed by the nonlinear finite element formulations. Additionally, the nonlinear finite element analysis procedures are also introduced in this chapter.

To consider bond–slip behaviour in between reinforcing bars and the surrounding concrete in numerical simulation, a composite beam element with bond–slip effect is introduced in Chapter 5. The displacement shape functions and the basic finite element formulations of the composite beam element are introduced. The material models, including bond stress–slip models for steel and FRP rebars, are then discussed, followed by the nonlinear finite element formulations for this composite beam element. Numerical examples are provided at the end of this chapter to demonstrate the finite element analyses of steel-/FRP-reinforced concrete beams with the consideration of bond–slip behaviour.

Chapter 6 discusses the finite element modelling of the structural performance of steel-/FRP-reinforced concrete beams at elevated temperatures. A composite beam element that considers the temperature effect is introduced in this chapter. A 2-D nonlinear heat transfer analysis is carried out to describe the temperature distribution throughout the cross section of the composite beam. The temperature-dependent material models of concrete, steel, and FRP are also discussed.

In addition, the use of commercial finite element analysis software ANSYS and the modelling techniques are introduced in Chapter 7 for the numerical simulation of FRP-strengthened reinforced concrete beams under both static and cyclic loads. The appropriate finite elements and material models available in ANSYS are introduced for modelling concrete, steel, and FRP; in particular, the description of the interface between FRP laminate and concrete substrate is discussed.

In this book, the Fortran codes developed for the finite element analyses using the 1-D composite beam elements under both ambient and fire conditions are provided in the appendixes, as well as the ANSYS user subroutines developed for concrete, steel, and FRP under cyclic loads. The reader is also referred to the temperature-dependent material properties in the appendixes.

References

[1] H. Toutanji, G. Ortiz, The effect of surface preparation on the bond interface between FRP sheets and concrete members, Compos. Struct. 53 (4) (2001) 457–462.

[2] B. Benmokrane, O. Chaallal, R. Masmoudi, Flexural response of concrete beams reinforced with FRP reinforcing bars, ACI Struct. J. 93 (1) (1996) 46–55.

[3] A. Ayoub, F.C. Filippou, Mixed formulation of nonlinear steel-concrete composite beam element, J. Struct. Eng. 126 (3) (2000) 371–381.

[4] C.E. Bakis, L.C. Bank, V.L. Brown, E. Cosenza, J.F. Davalos, J.J. Lesko, A. Machida, S.H. Rizkalla, T.C. Triantafillou, Fiber-reinforced polymer composites for construction—state-of-the-art review, J. Compos. Constr. 6 (2) (2002) 73–87.
[5] S. Rizkalla, T. Hassan, N. Hassan, Design recommendations for the use of FRP for reinforcement and strengthening of concrete structures, Prog. Struct. Eng. Mater. 5 (1) (2003) 16–28.
[6] V.L. Brown, C.L. Bartholomew, FRP reinforcing bars in reinforced concrete members, ACI Mater. J. 90 (1) (1993) 34–39.
[7] J.F. Davalos, Y. Chen, I. Ray, Effect of FRP bar degradation on interface bond with high strength concrete, Cem. Concr. Compos. 30 (8) (2008) 722–730.
[8] E. Cosenza, G. Manfredi, R. Realfonzo, Behavior and modeling of bond of FRP rebars to concrete, J. Compos. Constr. 1 (2) (1997) 40–51.
[9] N. Plevris, T.C. Triantafillou, D. Veneziano, Reliability of RC members strengthened with CFRP laminates, J. Struct. Eng. 121 (7) (1995) 1037–1044.
[10] O. Buyukozturk, O. Gunes, E. Karaca, Progress on understanding debonding problems in reinforced concrete and steel members strengthened using FRP composites, Constr. Build. Mater. 18 (1) (2004) 9–19.
[11] M.J. Chajes, T.A. Thomson, T.F. Januszka, W.W. Finch, Flexural strengthening of concrete beams using externally bonded composite materials, Constr. Build. Mater. 8 (3) (1994) 191–201.
[12] M.A. Aiello, M. Leone, M. Pecce, Bond performances of FRP rebars-reinforced concrete, J. Mater. Civ. Eng. 19 (3) (2007) 205–213.
[13] R. Okelo, R.L. Yuan, Bond strength of fiber reinforced polymer rebars in normal strength concrete, J. Compos. Constr. 9 (3) (2005) 203–213.
[14] L.J. Malvar, J.V. Cox, K.B. Cochran, Bond between carbon fiber reinforced polymer bars and concrete. I. Experimental study, J. Compos. Constr. 7 (2) (2001) 154–163.
[15] L.A. Bisby, M.F. Green, V.K.R. Kodur, Response to fire of concrete structures that incorporate FRP, Prog. Struct. Eng. Mater. 7 (3) (2005) 136–149.
[16] M. Saafi, Effect of fire on FRP reinforced concrete members, Compos. Struct. 58 (1) (2002) 11–20.
[17] L.A. Bisby, V.R. Kodur, M.F. Green, Performance of fire of FRP-confined reinforced concrete columns, in: 4th International Conference on Advanced Composite Materials in Bridges and Structures, Calgary, Alberta, 2004, pp. 1–8.
[18] Y. Yun, Y.F. Wu, W.C. Tang, Performance of FRP bonding systems under fatigue loading, Eng. Struct. 30 (11) (2008) 3129–3140.
[19] Y.J. Kim, P.J. Heffernan, Fatigue behavior of externally strengthened concrete beams with fiber-reinforced polymers: state of the art, J. Compos. Constr. 12 (3) (2008) 246–256.
[20] P.J. Heffernan, M.A. Erki, Fatigue behavior of reinforced concrete beams strengthened with carbon fiber reinforced plastic laminates, J. Compos. Constr. 8 (2) (2004) 132–140.
[21] Z. Wu, X. Wang, K. Iwashita, T. Sasaki, Y. Hamaguchi, Tensile fatigue behaviour of FRP and hybrid FRP sheets, Compos. Part B 41 (5) (2010) 396–402.
[22] P.T. Curtis, The fatigue behaviour of fibrous composite materials, J. Strain Anal. Eng. Des. 24 (4) (1989) 235–244.
[23] A. Hrennikoff, Solution of problems of elasticity by the framework method, J. Appl. Mech. 8 (1941) 169–175.

Finite element analysis of beams

2

2.1 Beam theories

In the real world, all structural members are in three dimensions as they have length, width, and thickness. However, in the case of beam-like structures, as length is normally much greater than the other two dimensions, they are often treated as 1-D problems in structural analysis. In finite element analysis, a beam theory could provide an effective method of reducing 3-D problems to more tractable 1-D calculations. It is applicable to the bodies that are initially prismatic and have small thicknesses compared with their overall dimensions [1]. Historically, investigations into the beam bending problem have been based mainly on the elementary Euler–Bernoulli beam theory (CBT) and shear-deformable Timoshenko beam theory (TBT). In addition, a number of beam theories, such as the higher-order shear deformation theory (HSDT), trigonometric shear deformation theory (TSDT), and exact beam theory (EBT), have also been established and used to represent the kinematics of beam deformation. It should be noted that all these beam theories are applicable for beams with cross sections in different shapes. In this section, a brief review of these beam theories is presented.

The classical Euler–Bernoulli beam theory, which applies to slender and prismatic beams, is the most commonly used beam theory. The fundamental assumption of CBT is that plane sections that are initially perpendicular to the centroidal axis remain plane and perpendicular to that axis after deformation. The displacement field of a Euler–Bernoulli beam takes the form of

$$u(x,z) = -z\frac{dw_0}{dx} \tag{2.1a}$$

$$w(x,z) = w_0(x) \tag{2.1b}$$

where the x-coordinate is taken along the length of the beam, the z-coordinate along the depth of the beam, and the y-coordinate along the width of the beam. Thus, w_0 is the transverse deflection in the midplane ($z=0$) of the beam.

Fig. 2.1A shows an undeformed beam. The kinematics of the CBT is shown in Fig. 2.1B, which indicates the plane-section assumption and the normality assumption. The neglect of transverse shear and transverse normal effects is also implied, which means that the deformation is due entirely to bending and in-plane stretching [3]. Therefore, the CBT is strictly correct only in the case of pure bending when the bending moment is constant along the beam. However, in practice, CBT is also acceptable for other bending moment distributions provided the beam is slender, since the effect of shear strain on the axial stress and lateral deflection is insignificant [1]. The CBT has also been extended to the well-known classical laminated beam theory

Nonlinear Finite Element Analysis of Composite and Reinforced Concrete Beams
https://doi.org/10.1016/B978-0-12-816899-8.00002-X

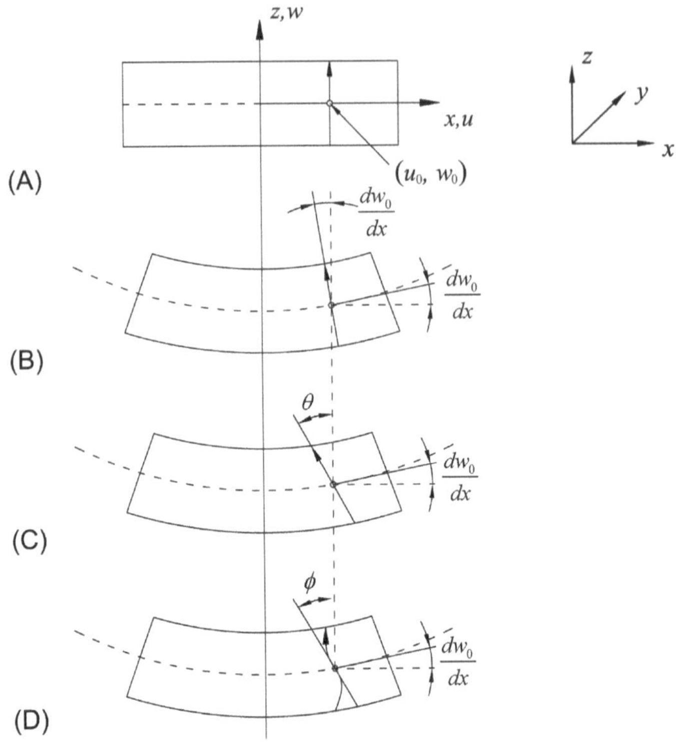

Fig. 2.1 Kinematics of Euler–Bernoulli, Timoshenko, and higher-order shear deformation beam theories [2].

(CLBT) for the analysis of laminated composite beams, and the effect of transverse shear deformation is neglected in CLBT as well.

When the depth of a beam is significant compared with its length, the effect of shear strain should not be neglected. Moreover, the influence of transverse shear deformation is much more pronounced for laminated composite beams than for isotropic beams due to the fact that most of the composite laminates have high in-plane Young's modulus to transverse shear modulus ratios [4]. In this case, the results produced by the CBT and CLBT are not accurate or correct.

To overcome the deficiency in CBT and CLBT, the shear deformation should be taken into account in the analysis of relatively short or deep beams. In the early 20th century, the Euler–Bernoulli equations were modified by Timoshenko [5], leading to a robust TBT, which is also known as the first-order shear deformation theory. It relaxes the restrictive assumption in CBT that a plane section remains plane after deformation but experiences a transverse shear strain. In Timoshenko's beam equations, the effective transverse shear strain is assumed to be constant with respect to the depth of the beam, which violates the 'zero transverse shear strain' boundary conditions on the top and bottom surfaces of the beam. Therefore, a nondimensional shear correction factor δ is introduced to account for the actual shear stress distribution that is quadratic along the depth of the beam. The transverse shear strain is then equal to the average shear stress of a cross section divided by the product of shear modulus and shear correction factor [6]. The displacement field of a Timoshenko beam is given by the following equations:

$$u(x,z) = -z\theta(x) \tag{2.2a}$$

$$w(x,z) = w_0(x) \tag{2.2b}$$

where θ denotes the rotation of a transverse normal about the y-axis rather than the rotation of the neutral axis, as shown in Fig. 2.1C.

To increase the accuracy of TBT, various shear correction factors have been proposed utilising different approaches and assumptions [6–8]. To date, the TBT has been widely used in finite element models, and a number of elements for isotropic or composite beams have been developed based on it [9–18].

On the other hand, the HSDT is usually considered to be able to better reproduce the kinematic behaviour of beams than the FSDT. In higher-order theories, the Euler–Bernoulli hypothesis is further relaxed by removing the plane-section assumption through introducing higher-order terms in the kinematic description of the displacement field. For example, the displacement field of a second-order beam theory with transverse inextensibility is given as follows [2]:

$$u(x,z) = -z\phi(x) + z^2\psi(x) \tag{2.3a}$$

$$w(x,z) = w_0(x) \tag{2.3b}$$

where ϕ represents the slope du/dx in the midplane ($z=0$) of the beam as can be seen in Fig. 2.1D and ϕ and ψ together define the quadratic nature of the deformed line.

Similarly, the displacement field of a third-order beam theory is given as [2]

$$u(x,z) = -z\phi(x) + z^2\psi(x) + z^3\xi(x) \tag{2.4a}$$

$$w(x,z) = w_0(x) \tag{2.4b}$$

where ξ is the unknown function of position x.

The displacement field of a third-order beam theory, which accommodates quadratic variations in the transverse shear strain and stresses, used by Levinson [19], Bickford [20], and Reddy [21] is expressed as Eqs (2.5a), (2.5b). The zero transverse shear stress at both top and bottom surfaces of the beam is satisfied, and no shear correction factor is required:

$$u(x,z) = -z\phi(x) + \alpha z^3\left(\phi + \frac{dw_0}{dx}\right) \tag{2.5a}$$

$$w(x,z) = w_0(x) \tag{2.5b}$$

where $\alpha = 4/(3h^2)$ and h is the thickness of the beam.

In addition to the aforementioned second-order and third-order beam theories, a large number of beam elements based on the HSDT for the analysis of isotropic or composite beams have also appeared in the literatures [3, 22–30]. They differ from each other in the choice of either interpolation functions used for the transverse deflection and rotation or the weak form used to develop the finite element model [3]. These higher-order beam elements take into account the parabolic distribution of the transverse shear strain and the stress-free conditions on the top and bottom surfaces of the beam; thus, the shear coefficient is omitted in these elements.

Apart from the CBT, TBT or FSDT, and HSDT, the TSDT, which contains the trigonometric sine, cosine, or hyperbolic sine and cosine functions in the thickness coordinate, is another beam theory that can be used to analyse beams. The trigonometric functions involving the thickness coordinate are associated with the transverse shear deformation effects and the shear stress distribution throughout the thickness of the beam. In the simplest form of TSDT, the number of displacement variables can be the same as those in the FSDT [31]. The application of TSDT has also been found in the studies of isotropic and composite beams [31–33]. Eqs (2.6a), (2.6b) show the displacement field proposed by Vlasov and Leont'ev [32], which introduce trigonometric functions into kinematic assumptions:

$$u(x,z) = zu_1(x) + u_2(x)\sin\left(\frac{2\pi z}{h}\right) \tag{2.6a}$$

$$w(x,z) = w_1(x) + v_2(x)\cos\left(\frac{\pi z}{h}\right) \tag{2.6b}$$

where the first terms on the right-hand sides represent displacements when the cross section is assumed to remain plane and the second terms are introduced to correct inaccuracies due to the plane-section assumption and that of zero transverse elongation. However, in this theory, the shear stress-free boundary conditions are not satisfied on the top and bottom surfaces of the beam [31].

The beam theory developed by Ghugal and Shimpi [31] based on the TSDT satisfies the zero transverse shear stress conditions on the boundaries and obviates the need to use the shear correction factor. The displacement field of this beam theory takes the form of

$$u(x,z) = u_0(x) - z\frac{dw_0}{dx} + \frac{h}{\pi}\sin\left(\frac{\pi z}{h}\right)\phi(x) \tag{2.7a}$$

$$w(x,z) = w_0(x) \tag{2.7b}$$

Although the TSDT seems to be a promising theory for solving beam problems, refined theories based on trigonometric functions are still not fully explored and need critical evaluation [34].

Besides, unlike the CBT, the EBT proposed by Ladevèze and Simmonds [35] for elastic straight beams is not limited to the relatively slender beams. In EBT, the 3-D solution of an equilibrium elasticity beam problem is the summation of a long-wavelength part, called the Saint-Venant solution, and a short-wavelength part, called the localised solution [36]. The solution s to the 3-D reference elasticity problem is given by

$$s(x) = s_{sv}(x) + \int_0^L s^{\pm}(x - t, t)dt \quad (2.8)$$

where s_{sv} is the long-wavelength part of the solution, $s^{\pm}$ a sort of Green function, and $t \rightarrow s^{\pm}(x - t, t)$ a local effect density that includes a regular part and Dirac distributions centred at both the abscissa of different discontinuities and the extremities.

The EBT has been implemented in several studies [37, 38]. However, as it requires large numbers of nodes and relations imposed between the degrees of freedom (DOFs), finite element models based on the EBT are much more complicated and inefficient. More importantly, the methods developed based on the EBT are not suited to the immediate use of standard finite element elasticity software due to the fact that the rigidity matrices they obtain are different from those of usual elasticity problems [36].

From the earlier discussion, the FSDT and HSDT seem to be the better alternatives to the other beam theories. However, one disadvantage of the HSDT compared with the TBT/FSDT is the presence of the higher-order derivative of the transverse deflection, which usually requires the development of a C^1 element [26], thus complicating the specifications of the boundary conditions. Also, the additional unknowns introduced by the higher-order theories are often difficult to interpret in physical terms, and the accuracy gained by using the theories higher than third order is so small that the effort required to solve the equations is not justified [2]. In fact, the plane-section assumption in the CBT and TBT is sufficiently accurate for isotropic beams and layered composite beams with plies of similar stiffness [39]. Therefore, the FSDT is still the most attractive approach due to its simplicity and low computational cost, and the global response predicted by the FSDT is fairly good.

2.2 Finite element analysis of beams

Finite element method is known as a robust numerical technique for finding approximate solutions to practical engineering problems, especially those with complicated geometries, boundary conditions, and loading conditions. It not only is a reliable tool for structural analysis but also is economical in terms of both resources and time. Although finite element method is versatile compared with other research techniques, difficulties may still be encountered. One of the most troubling problems in the finite element simulation of beams is that inaccurate results often arise when using the same element to model beams with different slenderness ratios; this is recognised as 'shear locking'. Another major obstacle is modelling beams made from various materials and suffering from material nonlinearity. Given these two problems, the finite element modelling techniques for solving them are discussed in the following sections.

2.2.1 Shear-locking phenomenon

Shear locking is a notorious phenomenon in the finite element analyses of beams and plates when elements used to analyse deep beams or thick plates are utilised to analyse slender beams or thin plates. This is characterised as the inability of the element to yield zero shear strains as it becomes progressively slender or thinner, thus leaving the deformation considerably underestimated [16]. Usually, the shear-locking phenomenon appears in the development of C^0 continuous elements using shear-flexible or penalty-type formulations. When a penalty-type argument is used to degenerate thick elements to thin elements, various approximations of the shear-related energy terms would act as different types of constraints. Depending on the formulation, two types of constraints, classified as true and spurious, may emerge, and the spurious constraints are responsible for the 'shear-locking' phenomenon [25]. In this case, overstiff solutions are always encountered due to the development of spurious stiffness contributions in the numerical solutions [40].

Up to now, many locking removal techniques have been developed and employed by researchers to eliminate shear locking in the analyses of slender beams and thin plates. The reduced integration technique, which facilitates the achievement of this aim, is one of the most commonly used methods. Exact numerical integration will preserve all the constraints, including both true and spurious constraints, whilst reduced integration can eliminate some or all of the spurious constraints and may eliminate some of the true constraints depending on the order of integration used [25]. This technique includes uniform scheme and selective scheme. In uniform reduced integration, both the bending and shear terms are integrated with the same rule that is of a lower order than the normal one [41]. This technique has been adopted by several investigators [41–44] and has shown considerable improvement in the capability of modelling. In selective reduced integration, the bending term is integrated with the normal rule, whereas the shear term is integrated with a lower-order rule [41]. Applications of the selective reduced integration method can also be found in quite a number of studies [41, 45–48]. However, a rank deficiency of the element stiffness matrix can be caused by using the reduced integration method, and thus, a kinematic deformation mode or zero-energy mode is frequently developed, which leads to an oscillation in the results [49, 50]. The reduced integration techniques have been proven to be generally insufficient for the development of robust elements [51].

An alternative procedure to the reduced integration method is the use of the constrained substitute shear strain field technique [51–53] that imposes 'a priori' adequately constrained shear strain fields, thereby allowing natural satisfaction of the conditions of vanishing transverse shear strains for a thin element limit. Although this approach has been proven to be successful in the literature, it is not easy to follow.

The discrete Kirchhoff technique has been utilised in the development of some plate elements [54–57], in which the classical equations are abandoned in favour of a bending theory that includes shear deformations. To capture the behaviour of thin plate, the constraint of zero shear strains is imposed at a discrete number of points. This method is effective, but its implementation tends to be somewhat complicated [58].

Moreover, in the mixed interpolation of tensorial components (MITC) family elements developed by Bathe and coworkers [59–63], shear strains are not evaluated from displacements; instead, separate interpolations are introduced for those strain components to solve the shear-locking problem. However, the difficulty of choosing proper orders of interpolation for out-of-plane displacements, section rotations, and transverse shear strains that result in nonlocking behaviour makes this technique not easy to handle.

In addition to the aforementioned techniques, some other methods for avoiding shear-locking issue have also been proposed. Raveendranath et al. [64] developed a shear-flexible beam element based on the coupled displacement field interpolations that share some of the generalised DOFs. The discrete shear gap (DSG) method proposed by Bletzinger et al. [65] is based on the explicit satisfaction of the kinematic equations for the shear strains at discrete points and effectively eliminates the parasitic shear strains. In the locking-free strain-based finite element formulation presented by Schnabl et al. [66], a modified principle of virtual work is employed as a basis for the finite element discretisation. Linear kinematic equations are included into the principle, and a strain field vector remains the only unknown function to be approximated in the finite element implementation of the principle. Although these methods are able to eliminate or alleviate shear locking, the accuracy of the finite elements may be reduced.

On the other hand, the formulation of interpolation functions that embody interdependency between bending and shear deformation appears to overcome the shear-locking phenomenon and results in an efficient and accurate locking-free finite element [3, 16]. This method has been employed in a number of studies for the development of unified formulations for finite elements over a wide range of thickness/length ratios [3, 16, 24], amongst which Timoshenko's beam function method has been proven to be simple and efficient for tackling shear-locking problem in plate elements, such as the triangular plate elements [67, 68], the quadrilateral plate element [69], the layered triangular plate element [70], and the layered quadrilateral plate elements [71–73]. Details of the Timoshenko's beam functions are discussed in the following section.

2.2.2 Timoshenko's beam functions

In the previous section, a number of locking removal techniques for avoiding the parasitic shear in bending elements have been discussed, and the majority of which have no direct physical interpretation and seem rather to be numerical tricks. In Timoshenko's beam function method, instead of considering the element stiffness matrix only as many other locking removal techniques do, the global expression of the equilibrium equation for a whole beam is analysed [12]. A linearly elastic Timoshenko beam element with four DOFs is shown in Fig. 2.2.

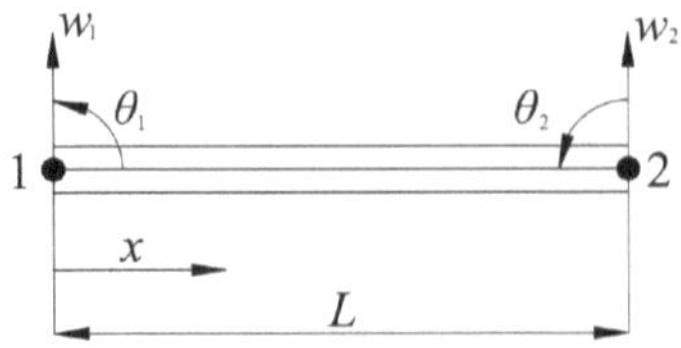

Fig. 2.2 Description of a Timoshenko beam element.

The interdependent cubic and quadratic polynomials are used to describe the transverse and rotational displacements, respectively [74]. The general shape functions of the transverse displacement w and rotation θ are assumed as

$$w(x) = \left\{ 1 \ \frac{x}{L} \ \left(\frac{x}{L}\right)^2 \ \left(\frac{x}{L}\right)^3 \right\} \begin{bmatrix} 1 & 0 & 0 & 0 \\ a_{11} & a_{21} & a_{31} & a_{41} \\ a_{12} & a_{22} & a_{32} & a_{42} \\ a_{13} & a_{23} & a_{33} & a_{43} \end{bmatrix} \begin{Bmatrix} w_1 \\ \theta_1 \\ w_2 \\ \theta_2 \end{Bmatrix} \quad (2.9a)$$

$$\theta(x) = \left\{ 1 \ \frac{x}{L} \ \left(\frac{x}{L}\right)^2 \right\} \begin{bmatrix} 0 & 1 & 0 & 0 \\ b_{11} & b_{21} & b_{31} & b_{41} \\ b_{12} & b_{22} & b_{32} & b_{42} \end{bmatrix} \begin{Bmatrix} w_1 \\ \theta_1 \\ w_2 \\ \theta_2 \end{Bmatrix} \quad (2.9b)$$

where w_1, θ_1, w_2, and θ_2 are the nodal displacements and rotations at beam end nodes 1 and 2, respectively, and a_{ij} and b_{ij} are unknown coefficients.

The shape functions are developed to exactly satisfy the homogeneous forms of static equations of equilibrium of an unstressed uniform Timoshenko beam. Thus, the coefficients in the assumed shape functions can be determined by substituting the shape functions into Eqs (2.10a), (2.10b) [74]:

$$\frac{\partial}{\partial x}\left(\delta GA\left(\frac{\partial w}{\partial x} - \theta\right)\right) = 0 \quad (2.10a)$$

$$\frac{\partial}{\partial x}\left(EI\frac{\partial \theta}{\partial x}\right) + \delta GA\left(\frac{\partial w}{\partial x} - \theta\right) = 0 \quad (2.10b)$$

where EI and δGA are the flexural rigidity and shear rigidity, respectively.

Consequently, the exact shape functions obtained for the two-node Timoshenko beam element are given in Table 2.1.

where α is the ratio of the beam bending stiffness to the shear stiffness, given by

$$\alpha = \frac{12}{L^2}\left(\frac{EI}{\delta GA}\right) \quad (2.11)$$

It is interesting to note that, for long slender beams, $\alpha \approx 0$; thus, one obtains the standard Hermitian polynomials for the classical Euler–Bernoulli beam, which is free of

Table 2.1 Exact shape functions for Timoshenko beam element.

	$w(x)$	$\theta(x)$
w_1	$\frac{1}{(1+\alpha)}\left\{2\left(\frac{x}{L}\right)^3-3\left(\frac{x}{L}\right)^2-\alpha\left(\frac{x}{L}\right)+(1+\alpha)\right\}$	$\frac{6}{(1+\alpha)L}\left\{\left(\frac{x}{L}\right)^2-\left(\frac{x}{L}\right)\right\}$
θ_1	$\frac{L}{(1+\alpha)}\left\{\left(\frac{x}{L}\right)^3-\left(2+\frac{\alpha}{2}\right)\left(\frac{x}{L}\right)^2+\left(1+\frac{\alpha}{2}\right)\left(\frac{x}{L}\right)\right\}$	$\frac{1}{(1+\alpha)}\left\{3\left(\frac{x}{L}\right)^2-(4+\alpha)\left(\frac{x}{L}\right)+(1+\alpha)\right\}$
w_2	$-\frac{1}{(1+\alpha)}\left\{2\left(\frac{x}{L}\right)^3-3\left(\frac{x}{L}\right)^2-\alpha\left(\frac{x}{L}\right)\right\}$	$-\frac{6}{(1+\alpha)L}\left\{\left(\frac{x}{L}\right)^2-\left(\frac{x}{L}\right)\right\}$
θ_2	$\frac{L}{(1+\alpha)}\left\{\left(\frac{x}{L}\right)^3-\left(1-\frac{\alpha}{2}\right)\left(\frac{x}{L}\right)^2-\frac{\alpha}{2}\left(\frac{x}{L}\right)\right\}$	$\frac{1}{(1+\alpha)}\left\{3\left(\frac{x}{L}\right)^2-(2-\alpha)\left(\frac{x}{L}\right)\right\}$

shear locking. The mass, incremental stiffness, and consistent force matrices can be reduced to the classical Euler–Bernoulli-based form by setting $\alpha=0$.

Timoshenko's beam function method has been used to construct the displacement shape functions for various plate elements as mentioned in the previous section, and by using the Timoshenko's beam functions, shear locking is avoided naturally without the need to employ any stabilisation schemes.

2.2.3 Layered method

When applying the finite element method to analyse structures with nonlinear material properties (material nonlinearity is discussed in Section 2.3.2), such as concrete structures, it is possible to model the cracking of concrete by allowing any node to split into two or more nodes, each capable of having independent displacement. A major disadvantage of this method is that it involves changes in topology as the solution proceeds. Other shortcomings of such a treatment are that the cracks can only propagate along element boundaries and the occurrence of cracking is defined from nodal stresses. This approach can be applied to 2-D and 3-D structures when they are idealised as such. However, if one of the dimensions is eliminated, this topology change approach should not be applied [75]. To simulate the material nonlinearity of a concrete structure, the layered approach is one of the most effective and efficient alternatives, which is based on the idealised stress–strain relationships of concrete and reinforcing materials [76]. To date, the layered approach has been widely and successfully used in many finite element analyses of reinforced concrete structures [77–81].

One of the earliest applications of the layered approach in reinforced concrete structures was reported by Hand et al. [75]. When a real 3-D problem is reduced to a 2-D plate or 1-D beam problem, the stress resultant concept is introduced. The layering concept can be considered as simply a discretised application of the stress resultant concept. By introducing the layered system, it is possible to have variations in material properties along the thickness of a beam or plate without resorting to a 2-D or 3-D finite element analysis. In the nonlayered approach, the stress resultants are integrated, in one step, over the total thickness and represent an average of the stress states and material properties. In this approach, there is usually only one set

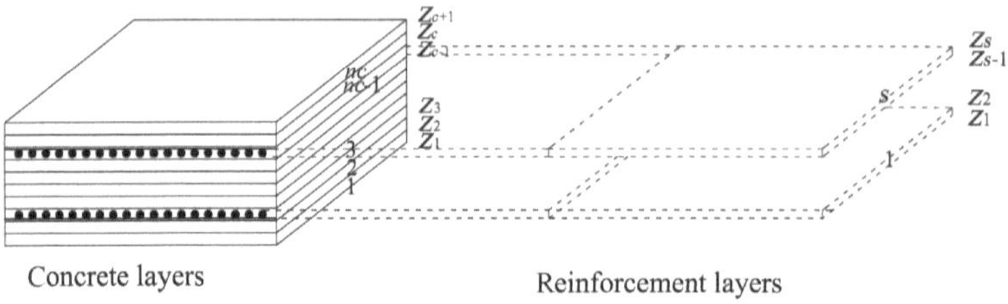

Fig. 2.3 Construction of a typical layered system.

of material properties allowed throughout the depth. In the layered approach, by contrast, the total thickness is divided into a number of layers. The stress resultants are defined for each layer separately by integrating them over only the thickness of that layer. The stress resultants for the total thickness are then obtained by summing those for the layers. As, for each layer, the method is independent of the specific types of material, the sum does not require all layers to have the same material properties. Thus, by employing the layered approach, a single finite element is able to include as many different material properties as the number of its layers, allowing variation in material properties throughout the thickness of the structure [75].

A typical layered structure is shown in Fig. 2.3. Each finite element is divided through the thickness into a chosen number of layers. Although each layer has a uniform thickness, the thickness may vary from layer to layer. Material models are then applied to each layer individually and assumed to be constant throughout the thickness of that layer. Therefore, different layers may have different material properties. The layers and interfaces are numbered sequentially from the bottom to the top of the element. For a reinforced concrete structure, its reinforcing bars are represented by smeared reinforcing layers with equivalent thicknesses. As, in the layered method, each layer is analysed individually, analyses of structures that have unusual cross-sectional shapes or reinforcing details are permitted. The only section compatibility requirement is that the plane sections remain plane so that the longitudinal strain in each layer can be fixed by defining the top and bottom fibre strains in the section [77].

In summary, by using the layered approach, the material properties of a particular layer that undergoes material degradation can be suitably modified, and the virgin material properties of the remaining layers of the element can be retained [82]. This approach can also account for the progressive growth of damage and plasticity throughout the thickness of a structure. The layered method has been demonstrated to be effective, particularly for predicting the cracking and ultimate behaviours of reinforced concrete beams and slabs [73].

2.3 Nonlinear finite element analysis of beams

In most engineering applications, the problems encountered are nonlinear. In general, nonlinearities in structural mechanics arise in two different ways: (1) from material or physical nonlinearity and (2) from geometric nonlinearity. In the case of material nonlinearity, the generalised Hooke's law is not valid if the stress–strain behaviour of a material is nonlinear, and the governing equations also become nonlinear. The effect of material nonlinearity on the behaviour of structural components is of great importance in finite element analysis. As for geometric nonlinearity, this type of nonlinearity is based on the deformations of an elastic body. When dealing with geometric nonlinearity, the relationships between the extensional strains and shear strains on the one hand and displacement components on the other are taken to be nonlinear, resulting in nonlinear strain–displacement relationships. In the case of beam problems, nonlinearity due to large deformations may be caused by the stretching of the neutral axis of the beam or by the nonlinear moment–curvature relationship [83].

2.3.1 Geometric nonlinearity

2.3.1.1 Geometric nonlinearity due to nonlinear moment–curvature relationship

For flexible beams undergoing large deformations as shown in Fig. 2.4, the nonlinear moment–curvature relationship is given by [83]

$$\frac{1}{R} = -\frac{w_{,xx}}{\left(1 + w_{,x}^2\right)^{3/2}} = \frac{M}{EI} \tag{2.12}$$

where R is the radius of curvature at a point on the deformed neutral axis of the beam, w the transverse displacement at any station x of the beam, E the modulus of elasticity, I the moment of inertia of the beam's cross section about the principal axis, and M the bending moment acting in a plane perpendicular to the axis of bending.

If the deformations are small, quantity $w_{,x}^2$ is negligible in comparison with unity, and thus, Eq. (2.12) reduces to a linear equation as

$$\frac{M}{EI} = -w_{,xx} \tag{2.13}$$

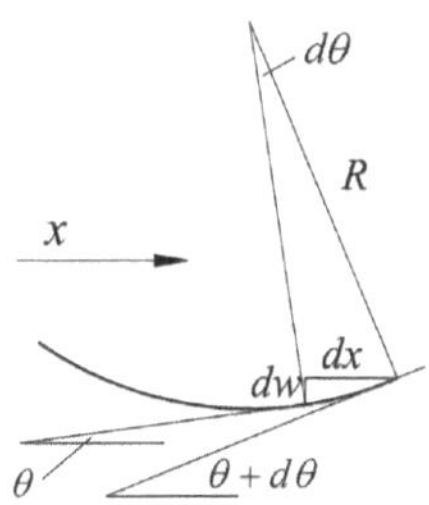

Fig. 2.4 A deformed beam element [83].

2.3.1.2 Geometric nonlinearity due to stretching

When a beam supported by unyielding supports undergoes large deflections, its neutral axis stretches, resulting in a nonlinear relationship between strain and displacement. Fig. 2.5 shows a beam element undergoing a large deformation. The axial strain ε_x is then given by [83]

$$\varepsilon_x = u_{,x} + \frac{1}{2}u_{,x}^2 + \frac{1}{2}w_{,x}^2 \approx u_{,x} + \frac{1}{2}w_{,x}^2 \tag{2.14}$$

For a beam made from an isotropic elastic material, the stresses may be written in terms of strains as

$$\sigma_x = E\varepsilon_x \tag{2.15}$$

The stress and moment resultants are then given by

$$N_x = \int \sigma_x dz \text{ and } M_x = \int \sigma_x z dz \tag{2.16}$$

2.3.2 Material nonlinearity

Another type of nonlinearity is due to the nonlinear behaviour of the material of a beam. There are very few materials that are linear over a wide range of strains and ideally plastic in the inelastic region [83]. It is universally acknowledged that reinforced concrete is a composite material with highly nonlinear material properties, since concrete exhibits an essentially nonlinear stress–strain relationship in compression under monotonic load and develops elastic–plastic properties with damaging effects under cyclic load whilst, in tension, crack formation and crack closure further add to its nonlinear material behaviour. In addition, steel reinforcement exhibits elastoplastic behaviour in both tension and compression, with a strain-hardening effect under cyclic load [84]. Hence, complicated nonlinear material laws, which are usually expressed by nonlinear response functions between stresses and strains or by a set of evolution equations, have to be considered to obtain realistic descriptions of engineering problems [85]. Furthermore, when subjected to elevated temperatures, the reduction in material properties of both concrete and reinforcing material may

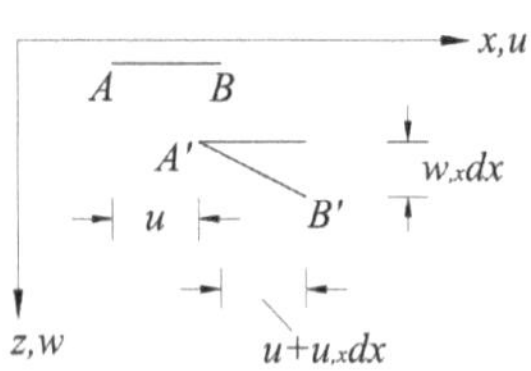

Fig. 2.5 Large deformation of a beam [83].

significantly affect the performance of a structure, especially FRP-reinforced concrete structures where the mechanical properties of FRP deteriorate rapidly at high temperatures. Therefore, temperature-dependent material properties are essential in carrying out the structural analysis at elevated temperatures.

2.3.3 Nonlinear finite element equations

Common forms of nonlinear equations that frequently arise in practice contain powers or products of unknowns, and the methods for solving them are intrinsically iterative in character [86]. At present, incremental methods, mathematical programming techniques, and finite difference and finite element methods have found application in the solutions to nonlinear static and dynamic problems. The finite element method has been widely accepted as an effective numerical technique for solving complex structural problems with different types of nonlinearities. Thus, nonlinear finite element equations are presented in this section.

A finite element analysis involving both geometric and material nonlinearities is usually performed most effectively by using an incremental formulation of the equations of motion, in which the variables are updated incrementally in correspondence with successive load steps. The incremental finite element equation that governs the response of the finite element system in a static analysis is given as [87]

$$[K_i]\{\Delta q_{i+1}\} = \{R_{i+1}\} - \{F_i\} \tag{2.17}$$

where $[K_i]$ is the tangential stiffness matrix corresponding to the configuration of the system in step i, $\{\Delta q_{i+1}\}$ the vector of nodal incremental displacements given by $\{\Delta q_{i+1}\} = \{q_{i+1}\} - \{q_i\}$, $\{R_{i+1}\}$ the vector of externally applied nodal loads corresponding to step $i+1$, and $\{F_i\}$ the vector of nodal forces corresponding to the internal element stresses in step i.

The basic equation to be solved in a nonlinear analysis is [87]

$$\{R_{i+1}(q^*)\} - \{F_{i+1}(q^*)\} = \{0\} \tag{2.18}$$

where q^* is the solution of the nodal displacements.

2.3.4 Solutions to nonlinear finite element equations

In the case of large deflection beam problems, several methods can be used to solve the nonlinear governing equations, such as the least-squares method, an iterative method, and a trial-and-error procedure, in which Newton–Raphson method is the one most commonly used for finding solutions to nonlinear finite element equations. In Newton–Raphson method, extrapolation procedure is utilised on the slope of the function at the guessed root [86].

Normally, a nonlinear system can be expressed as

$$\{f(q^*)\} = \{0\} \tag{2.19}$$

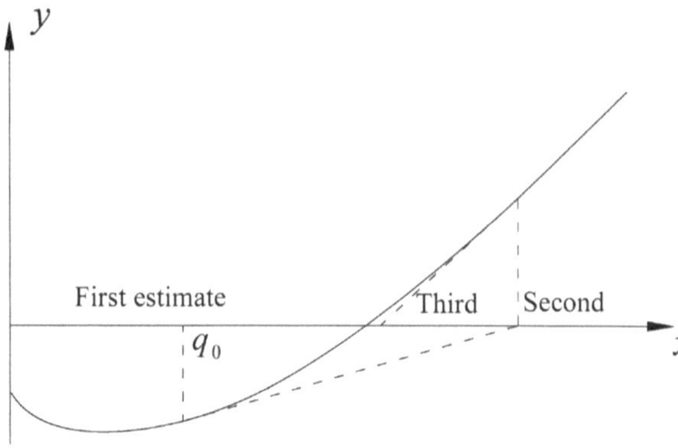

Fig. 2.6 Process of Newton–Raphson method [86].

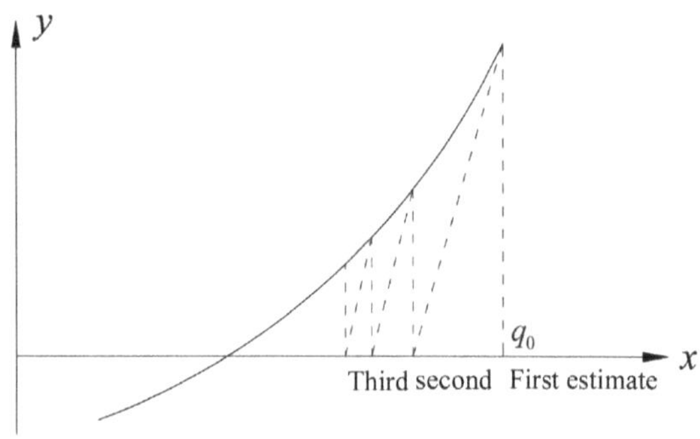

Fig. 2.7 Modified Newton–Raphson process with a constant gradient [86].

where

$$\{f(q^*)\} = \{R_{i+1}(q^*)\} - \{F_{i+1}(q^*)\} \tag{2.20}$$

The unknowns are the nodal values, and the equations represent a balance between the external and internal forces. There is a single unknown and a single equation for each DOF in the model. In incremental problems, instead of computing the displacements, their increments are computed. The external load is divided into a number of load steps ΔR and applied to the model step by step. In Newton–Raphson method, each iteration requires the computation and assembly of the stiffness matrix $[K_i]$ at the current state $\{q_i\}$, which is the matrix of the first derivatives $\partial\{f_i(q_i)\}/\partial\{q_i\}$, followed by the solution to the governing equation (Eq. 2.21) [88]. The process of Newton–Raphson method is illustrated in Fig. 2.6:

$$[K_i](\{q_{i+1}\} - \{q_i\}) = f(q_i) \tag{2.21}$$

The full Newton iteration is demonstrated to be efficient in many nonlinear structural analyses, whilst, for large systems of equations, the need to be forever calculating derivatives can make the iteration process expensive. In this case, a modified Newton–Raphson method can be used as an alternative, in which the first evaluation of $[K_0]$ is used for all further extrapolations. As can be seen in Fig. 2.7, the extrapolation lines are parallel to a gradient dictated by the derivative corresponding to the initial guessed value of q_0 [86].

However, the modified Newton–Raphson method using the tangential stiffness matrix $[K_0]$ computed at the start of the step runs into serious difficulties if the step brings large changes in the material properties. This has to be expected in plasticity because, when an elastic element becomes plastic, the convergence is slow and, when a plastic element becomes elastic, the divergence can be very fast [88].

References

[1] R.J. Astley, Finite Elements in Solids and Structures: An Introduction, first ed., Chapman & Hall, London, 1992.
[2] C.M. Wang, J.N. Reddy, K.H. Lee, Shear Deformable Beams and Plates, first ed., Elsevier Science Ltd., Oxford, 2000.
[3] J.N. Reddy, On locking-free shear deformable beam finite elements, Comput. Methods Appl. Mech. Eng. 149 (1–4) (1997) 113–132.
[4] T. Kant, K. Swaminathan, Analytical solutions for free vibration of laminated composite and sandwich plates based on a higher-order refined theory, Compos. Struct. 53 (1) (2001) 73–85.
[5] S.P. Timoshenko, On the correction for shear of the differential equation for transverse vibrations of prismatic bars, Philos. Mag. 41 (245) (1921) 744–746.
[6] G.R. Cowper, The shear coefficient in Timoshenko's beam theory, J. Appl. Mech. 33 (2) (1966) 335–340.
[7] C.W. Bert, Simplified analysis of static shear factors for beams of nonhomogeneous cross section, J. Compos. Mater. 7 (4) (1973) 525–529.
[8] N.G. Stephen, Considerations on second order beam theories, Int. J. Solids Struct. 17 (3) (1981) 325–333.
[9] K.K. Kapur, Vibrations of a Timoshenko beam, using finite-element approach, J. Acoust. Soc. Am. 40 (5) (1966) 1058–1063.
[10] R. Davis, R.D. Henshell, G.B. Warburton, A Timoshenko beam element, J. Sound Vib. 22 (4) (1972) 475–487.
[11] D.L. Thomas, J.M. Wilson, R.R. Wilson, Timoshenko beam finite elements, J. Sound Vib. 31 (3) (1973) 315–330.
[12] J. Rakowski, The interpretation of the shear locking in beam elements, Comput. Struct. 37 (5) (1990) 769–776.
[13] Z. Friedman, J.B. Kosmatka, An improved two-node timoshenko beam finite element, Comput. Struct. 47 (3) (1993) 473–481.
[14] M. Eisenberger, Derivation of shape functions for an exact 4-D.O.F. Timoshenko beam element, Commun. Numer. Methods Eng. 10 (9) (1994) 673–681.
[15] A. Bazoune, Y.A. Khulief, N.G. Stephen, Shape functions of three-dimensional Timoshenko beam element, J. Sound Vib. 259 (2) (2003) 473–480.
[16] I.B. Edem, The exact two-node Timoshenko beam finite element using analytical bending and shear rotation interdependent shape functions, Int. J. Computat. Methods Eng. Sci. Mech. 7 (6) (2006) 425–431.
[17] Y. Kim, J.F. Davalos, E.J. Barbero, Composite beam element with layerwise plane sections, J. Eng. Mech. 120 (5) (1994) 1160–1166.
[18] A.T. Chen, T.Y. Yang, Static and dynamic formulation of a symmetrically laminated beam finite element for a microcomputer, J. Compos. Mater. 19 (5) (1985) 459–475.
[19] M. Levinson, A new rectangular beam theory, J. Sound Vib. 74 (1) (1981) 81–87.

[20] W.B. Bickford, A consistent higher order beam theory, Dev. Theor. Appl. Mech. 11 (1982) 137–150.
[21] J.N. Reddy, A simple higher-order theory for laminated composite plates, J. Appl. Mech. 51 (4) (1984) 745–752.
[22] R.E. Nickel, G.A. Secor, Convergence of consistently derived Timoshenko beam finite elements, Int. J. Numer. Methods Eng. 5 (2) (1972) 243–252.
[23] D.J. Dawe, A finite element for the vibration analysis of Timoshenko beams, J. Sound Vib. 60 (1) (1978) 11–20.
[24] A. Tessler, S.B. Dong, On a hierarchy of conforming Timoshenko beam elements, Comput. Struct. 14 (3–4) (1981) 335–344.
[25] G. Prathap, G.R. Bhashyam, Reduced integration and the shear-flexible beam element, Int. J. Numer. Methods Eng. 18 (2) (1982) 195–210.
[26] P.R. Heyliger, J.N. Reddy, A higher order beam finite element for bending and vibration problems, J. Sound Vib. 126 (2) (1988) 309–326.
[27] I. Senjanović, Y. Fan, A higher-order flexural beam theory, Comput. Struct. 32 (5) (1989) 973–986.
[28] K.P. Soldatos, I. Elishakoff, A transverse shear and normal deformable orthotropic beam theory, J. Sound Vib. 155 (3) (1992) 528–533.
[29] S.R. Bhate, U.N. Nayak, A.V. Patki, Deformation of composite beam using refined theory, Comput. Struct. 54 (3) (1995) 541–546.
[30] K. Chandrashekhara, K.M. Bangera, Free vibration of composite beams using a refined shear flexible beam element, Comput. Struct. 43 (4) (1992) 719–727.
[31] Y.M. Ghugal, R.P. Shimpi, A review of refined shear deformation theories for isotropic and anisotropic laminated beams, J. Reinf. Plast. Compos. 20 (3) (2001) 255–272.
[32] V.Z. Vlasov, N.N. Leont'ev, Beams, Plates and Shells on Elastic Foundations, Israel Program for Scientific Translations Ltd, Jerusalem, 1966.
[33] R.P. Shimpi, A.V. Ainapure, A beam finite element based on layerwise trigonometric shear deformation theory, Compos. Struct. 53 (2) (2001) 153–162.
[34] Y.M. Ghugal, R.P. Shimpi, A review of refined shear deformation theories of isotropic and anisotropic laminated plates, J. Reinf. Plast. Compos. 21 (9) (2002) 775–813.
[35] P. Ladevèze, J. Simmonds, Newconcepts for linearbeamtheory with arbitrary geometry and loading, Eur. J. Mech. A/Solids 17 (3) (1998) 377–402.
[36] R.E. Fatmi, H. Zenzri, A numerical method for the exact elastic beam theory. Applications to homogeneous and composite beams, Int. J. Solids Struct. 41 (9–10) (2004) 2521–2537.
[37] P. Ladevèze, P. Sanchez, J.G. Simmonds, On application of the exact theory of elastic beams, Solid Mech. Appl. 88 (2002) 181–196.
[38] R.E. Fatmi, H. Zenzri, On the structural behavior and the Saint Venant solution in the exact beamtheory: Application to laminated composite beams, Comput. Struct. 80 (16–17) (2002) 1441–1456.
[39] J.F. Davalos, Y. Kim, E.J. Barbero, Analysis of laminated beams with a layer-wise constant shear theory, Compos. Struct. 28 (3) (1994) 241–253.
[40] G. Eccher, K.J.R. Rasmussen, R. Zandonini, Shear Locking in Isoparametric Spline Finite Strips, School of Civil Engineering, the University of Sydney, Sydney, 2007.
[41] T.J.R. Hughes, M. Cohen, M. Haroun, Reduced and selective integration techniques in the finite element analysis of plates, Nucl. Eng. Des. 46 (1) (1978) 203–222.
[42] O.C. Zienkiewicz, R.L. Taylor, J.M. Too, Reduced integration technique in general analysis of plates and shells, Int. J. Numer. Methods Eng. 3 (1971) 275–290.

[43] D. García-Vallejo, A.M. Mikkola, J.E.L. Escalona, A new locking-free shear deformable finite element based on absolute nodal coordinates, Nonlinear Dynam. 50 (1–2) (2007) 249–264.
[44] A.K. Noor, J.M. Peters, Mixed models and reduced/selective integration displacement models for nonlinear analysis of curved beams, Int. J. Numer. Methods Eng. 17 (4) (1981) 615–631.
[45] T.J.R. Hughes, R.L. Taylor, W. Kalcjai, Simple and efficient element for plate bending, Int. J. Numer. Methods Eng. 11 (10) (1977) 1529–1543.
[46] D.S. Malkus, T.J.R. Hughes, Mixed finite element methods-reduced and selective integration techniques: a unification of concepts, Comput. Methods Appl. Mech. Eng. 15 (1) (1978) 63–81.
[47] R.L. Taylor, F. Auricchio, Linked interpolation for Reissner–Mindlin plate elements. Part II. A simple triangle, Int. J. Numer. Methods Eng. 36 (18) (1993) 3057–3066.
[48] K.S. Kerkkänen, J.T. Sopanen, A.M. Mikkola, A linear beam finite element based on the absolute nodal coordinate formulation, J. Mech. Des. 127 (4) (2005) 621–630.
[49] Y. Long, F. Xi, A universal method for including shear deformation in thin plate elements, Int. J. Numer. Methods Eng. 34 (1) (1992) 171–177.
[50] F. Gruttmann, W. Wagner, A stabilized one-point integrated quadrilateral Reissner–Mindlin plate element, Int. J. Numer. Methods Eng. 61 (13) (2004) 2273–2295.
[51] E. Onate, O.C. Zienkiewicz, B. Surez, R.L. Taylor, A general methodology for deriving shear constrained Reissner–Mindlin plate elements, Int. J. Numer. Methods Eng. 33 (2) (1992) 345–367.
[52] E. Hinton, H.C. Huang, A family of quadrilateral Mindlin plate elements with substitute shear strain fields, Comput. Struct. 23 (3) (1986) 409–431.
[53] O.C. Zienkiewicz, R.L. Taylor, P. Papadopoulos, E. Onate, Plate bending elements with discrete constraints: new triangular elements, Comput. Struct. 35 (4) (1990) 505–522.
[54] J.L. Batoz, P. Lardeur, A discrete shear triangular nine DOF element for the analysis of thick to very thin plates, Int. J. Numer. Methods Eng. 29 (3) (1989) 533–560.
[55] J.L. Batoz, I. Katili, On a simple triangular Reissner/Mindlin plate element based on incompatible modes and discrete constraints, Int. J. Numer. Methods Eng. 35 (8) (1992) 1603–1632.
[56] I. Katili, A new discrete Kirchhoff-Mindlin element based on Mindlin–Reissner plate theory and assumed shear strain fields. Part I. An extended DKT element for thick-plate bending analysis, Int. J. Numer. Methods Eng. 36 (1993) 1859–1883.
[57] I. Katili, A new discrete Kirchhoff-Mindlin element based on Mindlin–Reissner plate theory and assumed shear strain fields. Part II. An extended DKQ element for thick-plate bending analysis, Int. J. Numer. Methods Eng. 36 (1993) 1885–1908.
[58] T.J.R. Hughes, R.L. Taylor, W. Kanoknukulchai, A simple and efficient finite element for plate bending, Int. J. Numer. Methods Eng. 11 (10) (1977) 1529–1543.
[59] E.N. Dvorkin, K.J. Bathe, A continuum mechanics based four-node shell element for general nonlinear analysis, Eng. Comput. 1 (1984) 77–88.
[60] K.J. Bathe, E.N. Dvorkin, A formulation of general shell elements - the use of mixed interpolation of tensorial components, Int. J. Numer. Methods Eng. 22 (3) (1986) 697–722.
[61] K.J. Bathe, E.N. Dvorkin, A four-node plate bending element based on Mindlin/Reissner plate theory and a mixed interpolation, Int. J. Numer. Methods Eng. 21 (2) (1985) 367–383.
[62] M.L. Bucalem, K.J. Bathe, Higher-order MITC general shell elements, Int. J. Numer. Methods Eng. 36 (21) (1993) 3729–3754.

[63] K.J. Bathe, A. Iosilevich, D. Chapelle, An evaluation of the MITC shell elements, Comput. Struct. 75 (1) (2000) 1–30.
[64] P. Raveendranath, G. Singh, G.V. Rao, A three-noded shear-flexible curved beam element based on coupled displacement field interpolations, Int. J. Numer. Methods Eng. 51 (1) (2001) 85–101.
[65] K.U. Bletzinger, M. Bischoff, E. Ramm, A unified approach for shear-locking-free triangular and rectangular shell finite elements, Comput. Struct. 75 (3) (2000) 321–334.
[66] S. Schnabl, M. Saje, G. Turk, I. Planinc, Locking-free two-layer Timoshenko beam element with interlayer slip, Finite Elem. Anal. Des. 43 (9) (2007) 705–714.
[67] A.K. Soh, Z.F. Long, S. Cen, A new nine DOF triangular element for analysis of thick and thin plates, Comput. Mech. 24 (5) (1999) 408–417.
[68] W. Chen, Y.K. Cheung, Refined 9-DOF triangular Mindlin plate elements, Int. J. Numer. Methods Eng. 51 (11) (2001) 1259–1281.
[69] W. Chen, Y.K. Cheung, Refined quadrilateral element based on Mindlin/Reissner plate theory, Int. J. Numer. Methods Eng. 47 (1–3) (2000) 605–627.
[70] Y.X. Zhang, K.S. Kim, A simple displacement-based 3-node triangular element for linear and geometrically nonlinear analysis of laminated composite plates, Comput. Methods Appl. Mech. Eng. 194 (2005) 4607–4632.
[71] S. Cen, Y. Long, Z. Yao, A new hybrid-enhanced displacement-based element for the analysis of laminated composite plates, Comput. Struct. 80 (9–10) (2002) 819–833.
[72] Y.X. Zhang, K.S. Kim, Two simple and efficient displacement-based quadrilateral elements for the analysis of composite laminated plates, Int. J. Numer. Methods Eng. 61 (11) (2004) 1771–1796.
[73] Y.X. Zhang, M.A. Bradford, R.I. Gilbert, A layered shear-flexural plate/shell element using Timoshenko beam functions for nonlinear analysis of reinforced concrete plates, Finite Elem. Anal. Des. 43 (2007) (2007) 888–900.
[74] J.B. Kosmatka, An improved two-node finite element for stability and natural frequencies of axial-loaded Timoshenko beams, Comput. Struct. 57 (1) (1995) 141–149.
[75] F.R. Hand, D.A. Pecknold, W.C. Schnobrich, A layered finite element nonlinear analysis of reinforced concrete plates and shells, in: Civil Engineering Studies SRS-389, 1972.
[76] C.K. Choi, H.G. Kwak, The effect of finite element mesh size in nonlinear analysis of reinforced concrete structures, Comput. Struct. 36 (5) (1990) 807–815.
[77] F.J. Vecchio, M.P. Collins, Predicting the response of reinforced concrete beams subjected to shear using modified compression field theory, ACI Struct. J. 85 (3) (1988) 258–268.
[78] Y.C. Loo, H. Guan, Cracking and punching shear failure analysis of RC flat plates, J. Struct. Eng. 123 (10) (1997) 1321–1330.
[79] R.S. Oliveira, M.A. Ramalho, M.R.S. Corrêa, A layered finite element for reinforced concrete beams with bond–slip effects, Cem. Concr. Compos. 30 (3) (2008) 245–252.
[80] A.J.M. Ferreira, P.P. Camanho, A.T. Marques, A.A. Fernandes, Modelling of concrete beams reinforced with FRP re-bars, Compos. Struct. 53 (2001) (2001) 107–116.
[81] D.R.J. Owen, J.A. Figueiras, F. Damjanic, Finite element analysis of reinforced and prestressed concrete structures including thermal loading, Comput. Methods Appl. Mech. Eng. 41 (3) (1983) 323–366.
[82] S. Chandrakanth, P.C. Pandey, Damage coupled elasto-plastic finite element analysis of a Timoshenko layered beam, Comput. Struct. 69 (3) (1998) 411–420.
[83] M. Sathyamoorthy, Nonlinear analysis of structures, in: F.A. Kulacki (Ed.), CRC Mechanical Engineering Series, CRC Press, New York, 1998.

[84] S.Y. Noh, W.B. Krätzig, K. Meskouris, Numerical simulation of serviceability, damage evolution and failure of reinforcedconcrete shells, Comput. Struct. 81 (8–11) (2003) 843–857.
[85] P. Wriggers, Nonlinear Finite Element Methods, Springer, Heidelberg, 2008.
[86] D.V. Griffiths, I.M. Smith, Numerical Methods for Engineers, second ed., Taylor & Francis Group, Boca Raton, 2006.
[87] K.J. Bathe, A.P. Cimento, Some practical procedures for the solution of nonlinear finite element equations, Comput. Methods Appl. Mech. Eng. 22 (1) (1980) 59–85.
[88] H. Matthies, G. Strang, The solution of nonlinear finite element equations, Int. J. Numer. Methods Eng. 14 (11) (1979) 1613–1626.

Finite element analysis of composite beams

3

3.1 Introduction

Over the last few decades, the advanced composite materials have gained worldwide interests due to their superior material performances, structural efficiencies, and cost effectiveness. The applications of composite materials can be found in various areas of science and technology, in particular in aerospace, mechanical, and civil and infrastructural engineering. Sandwich or laminated composite beam as shown in Fig. 3.1 is one of the most commonly encountered structural components.

Finite element method is universally acknowledged as a powerful tool for solving engineering problems. It has been widely and successfully applied in the simulation of behaviours of engineering structures, including composite structures such as laminated beams and plates. The employment of simple and efficient element is most crucial for effective finite element analysis. To study and predict the structural behaviour of composite beams, various types of finite elements have been developed, including one-dimensional (1-D), two-dimensional (2-D), and three-dimensional (3D) elements.

2-D and 3-D elements have often been used in the previous studies for finite element analysis of composite beams. For example, Ferreira et al. [1] modelled FRP-reinforced concrete composite beams using degenerated 2-D shell elements based on the first-order shear deformation theory. Yu [2] proposed a six-noded higher-order triangular layered shell element with six degrees of freedom at each node for the analysis of composite laminated plates and beams. In the study conducted by Manjunatha and Kant [3], 3-D discrete Lagrangian four-noded cubic elements with five, six, and seven degrees of freedoms per node were applied to the numerical analyses of symmetric/unsymmetric composite and sandwich beams. In addition, a generalised cross-sectional modelling approach for composite beams was utilised by Neto et al. [4] and Yu et al. [5], in which the complex 3-D elasticity problem was asymptotically decoupled into a linear 2-D cross-sectional analysis and a nonlinear 1-D beam problem.

Although 2-D and 3-D elements are usually able to produce more precise and accurate numerical predictions than 1-D elements, they are much more complicated in terms of both formulation and modelling. Also, 2-D and 3-D elements may cost far more computational space and resources due to the large numbers of nodes and degrees of freedom. By contrast, 1-D element is more computationally economic and efficient for analysis of beam-like structures. A number of 1-D elements have been developed for the analysis of composite beams, but most are based on the higher-order shear deformation theory with a large number of degrees of freedom. For example, Yuan and Miller [6] proposed a 1-D laminated beam element with five nodes and 16 degrees of freedom, Subramanian [7, 8] developed a two-noded beam

Nonlinear Finite Element Analysis of Composite and Reinforced Concrete Beams
https://doi.org/10.1016/B978-0-12-816899-8.00003-1

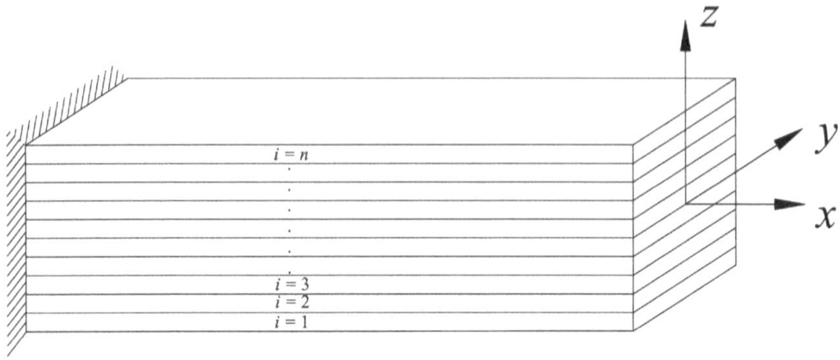

Fig. 3.1 Laminated composite beam.

element with eight degrees of freedom per node, a two-noded beam element with four degrees of freedom at each node was refined by Murthy et al. [9], and a quasi-conforming beam element with two nodes and four degrees of freedom per node was developed by Shi et al. [10]. In addition, a 1-D layer-wise constant shear beam element was developed by Davalos et al. [11] based on the generalised laminated plate theory, the first-order Hermite interpolation polynomials was adopted by Murthy and Rao [12] to construct displacement field for a laminated anisotropic beam element with 16 degrees of freedom, and Argyris et al. [13] proposed a two-noded composite shear-deformable beam element with 12 degrees of freedom.

Recently, the authors have developed a simple 1-D composite beam element based on the first-order shear deformation theory [14]. The element has only two nodes and two degrees of freedom per node, which is simple in both geometry and formulation, and computationally efficient for the analysis of composite beams. The Timoshenko's beam functions are employed to construct the displacement shape functions of the beam element, providing a unified formulation for both slender and deep beam analyses, and the notorious shear-locking problem is avoided naturally without the employment of any remedial scheme. A layered approach is utilised to describe the layered characteristics of composite beams. In this chapter, the element is introduced in details, and then, it is employed for the finite element analysis of composite beams. The procedures of finite element analyses of homogeneous and composite beams are also presented in details.

3.2 A one-dimensional two-node composite beam element

The composite beam element with two nodes and two degrees of freedom at each node (transverse displacement w and rotation θ) and its cross section are shown in Fig. 3.2 [14]. The cross section is composed of a number of layers that are numbered sequentially. The material properties of each layer are determined according to the layups of

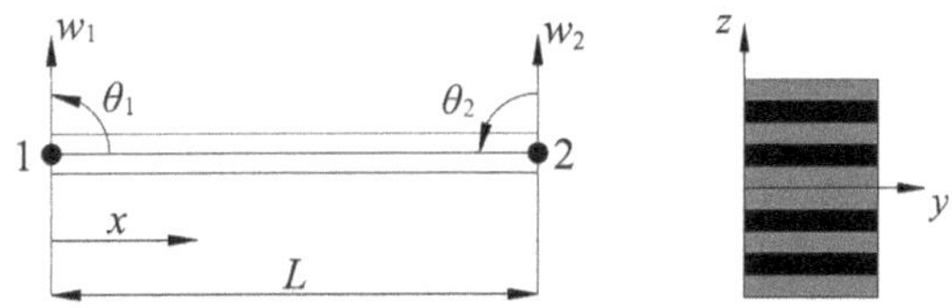

Fig. 3.2 A two-node composite beam element and its cross section [14].

the composite beams. Specially, for the case of homogeneous beams, the material properties are the same for each layer.

It is assumed that each layer is in a state of plane stress and the material properties are constant throughout the thickness of each layer. The material properties of the whole cross section are then obtained by algebraically summing the contribution of each layer as

$$D_{bb} = \frac{1}{3}\sum_{i=1}^{n} b_i E_i \left(z_{i+1}^3 - z_i^3\right) \tag{3.1}$$

$$D_{ss} = \delta \sum_{i=1}^{n} b_i G_i (z_{i+1} - z_i) \tag{3.2}$$

where D_{bb} and D_{ss} are the bending stiffness and transverse shear stiffness, respectively; b_i the width of the ith layer of the beam element; n the number of layers; δ the shear correction factor that is a constant of the nonuniformity of the shearing stress. The value of δ can be obtained based on different theories [15]. It is set to 5/6 in this book.

E_i is the bending elastic modulus of the ith layer; G_i the shear modulus of the ith layer; and z_{i+1} and z_i the coordinates of the upper and lower surfaces of the ith layer in z direction, respectively.

3.2.1 Basic formulations

Assuming that the in-plane forces are zero and the in-plane displacements (u_0 v_0) in the midplane are zero, the displacement field of the beam element takes the form of

$$u(x,z) = -z\theta(x) \tag{3.3}$$

$$w(x,z) = w_0(x) \tag{3.4}$$

where w_0 is the transverse displacement in the midplane and θ the rotation of the beam element.

The strain–displacement relationships are

$$\varepsilon_x = \frac{\partial u}{\partial x} = -z\frac{d\theta}{dx} \tag{3.5}$$

$$\varepsilon_b = -\frac{d\theta}{dx} \tag{3.6}$$

$$\gamma = \frac{\partial u}{\partial z} + \frac{\partial w}{\partial x} = \frac{\partial w}{\partial x} - \theta \tag{3.7}$$

where ε_x is the longitudinal strain, ε_b the bending strain, γ the transverse shear strain, and $\partial w/\partial x$ is the slope of the neutral axis.

3.2.2 Displacement functions of the composite beam element

A notorious phenomenon of shear locking often occurs when an element for simulating moderately deep beams is used to model slender beams. In this composite beam element, the Timoshenko's beam functions are employed to construct the displacement shape functions, which could perfectly solve the shear-locking problem, providing a unified beam element for analyses of both slender and moderately deep beams. The Timoshenko's beam functions are taken in the form of Eqs (3.8), (3.9) [16, 17].

The rotations and transverse displacements at any point along the longitudinal direction of the beam element can be represented using the displacements and rotations at the two ends of the composite beam element (w_1, w_2, θ_1, θ_2). The formulas of deflection w and rotation θ for the composite beam element with length L, width b, and height h are given as

$$\begin{aligned} w = {} & (L_1 + \mu_e L_1 L_2 (L_1 - L_2)) w_1 + (L_1 L_2 + \mu_e L_1 L_2 (L_1 - L_2)) L/2 \theta_1 \\ & + (L_2 + \mu_e L_1 L_2 (L_2 - L_1)) w_2 + (-L_1 L_2 + \mu_e L_1 L_2 (L_1 - L_2)) L/2 \theta_2 \end{aligned} \tag{3.8}$$

$$\theta = -\left(\frac{6L_1L_2}{L}\right)\mu_e w_1 + L_1(1 - 3\mu_e L_2)\theta_1 + \left(\frac{6L_1L_2}{L}\right)\mu_e w_2 + L_2(1 - 3\mu_e L_1)\theta_2 \tag{3.9}$$

where

$$L_1 = 1 - \frac{x}{L} \tag{3.10a}$$

$$L_2 = \frac{x}{L} \tag{3.10b}$$

$$\mu_e = \frac{1}{1 + 12\lambda_e} \tag{3.10c}$$

$$\lambda_e = \frac{D_{bb}}{D_{ss} L^2} \tag{3.10d}$$

in which x is the coordinate along the beam element and L the length of the beam element. For a composite beam, its bending stiffness and transverse shear stiffness (D_{bb} and D_{ss}, respectively) are given by Eqs (3.1), (3.2).

Eqs (3.8), (3.9) can be rewritten in a simpler form as

$$w = N_1 w_1 + N_2 \theta_1 + N_3 w_2 + N_4 \theta_2 \tag{3.11}$$

$$\theta = N_5 w_1 + N_6 \theta_1 + N_7 w_2 + N_8 \theta_2 \tag{3.12}$$

where

$$N_1 = L_1 + \mu_e L_1 L_2 (L_1 - L_2) \tag{3.13a}$$

$$N_2 = (L_1 L_2 + \mu_e L_1 L_2 (L_1 - L_2)) L/2 \tag{3.13b}$$

$$N_3 = L_2 + \mu_e L_1 L_2 (L_2 - L_1) \tag{3.13c}$$

$$N_4 = (-L_1 L_2 + \mu_e L_1 L_2 (L_1 - L_2)) L/2 \tag{3.13d}$$

$$N_5 = -\left(\frac{6 L_1 L_2}{L}\right) \mu_e \tag{3.13e}$$

$$N_6 = L_1 (1 - 3\mu_e L_2) \tag{3.13f}$$

$$N_7 = \left(\frac{6 L_1 L_2}{L}\right) \mu_e \tag{3.13g}$$

$$N_8 = L_2 (1 - 3\mu_e L_1) \tag{3.13h}$$

Hence, the transverse displacement and rotation can be expressed in matrix form as

$$\begin{Bmatrix} w \\ \theta \end{Bmatrix} = \begin{bmatrix} N_1 & N_2 & N_3 & N_4 \\ N_5 & N_6 & N_7 & N_8 \end{bmatrix} \begin{Bmatrix} w_1 \\ \theta_1 \\ w_2 \\ \theta_2 \end{Bmatrix} \tag{3.14}$$

Therefore, the shape function matrix $[N]$ is defined as

$$[N] = \begin{bmatrix} N_1 & N_2 & N_3 & N_4 \\ N_5 & N_6 & N_7 & N_8 \end{bmatrix} \tag{3.15}$$

3.2.3 Strain and strain matrix

The element's bending strain and shear strain can be expressed as

$$\varepsilon_b = [B_b] \left\{ q^{(e)} \right\} \tag{3.16}$$

$$\gamma = [B_s] \left\{ q^{(e)} \right\} \tag{3.17}$$

in which

$$[B_b] = -\left[\frac{dN_5}{dx}\ \frac{dN_6}{dx}\ \frac{dN_7}{dx}\ \frac{dN_8}{dx}\right]$$
$$= \left[\frac{6\mu_e}{L}\left(\frac{1}{L}-\frac{2x}{L^2}\right)\ \frac{3\mu_e+1}{L} - 6\mu_e\frac{x}{L^2}\ -\frac{6\mu_e}{L}\left(\frac{1}{L}-\frac{2x}{L^2}\right)\ -\frac{1-3\mu_e}{L} - 6\mu_e\frac{x}{L^2}\right] \tag{3.18a}$$

$$[B_s] = \left[\frac{dN_1}{dx} - N_5\ \ \frac{dN_2}{dx} - N_6\ \ \frac{dN_3}{dx} - N_7\ \ \frac{dN_4}{dx} - N_8\right]$$
$$= \left[-\frac{1}{L}+\frac{\mu_e}{L}\ \ \frac{\mu_e}{2}-\frac{1}{2}\ \ \frac{1}{L}-\frac{\mu_e}{L}\ \ -\frac{1}{2}+\frac{\mu_e}{2}\right] \tag{3.18b}$$

where $[B_b]$ is the bending strain matrix, $[B_s]$ the shear strain matrix, and $\{q^{(e)}\}$ the element nodal displacement vector that can be expressed as

$$\left\{q^{(e)}\right\} = \begin{Bmatrix} q_1 \\ q_2 \end{Bmatrix} \text{ with } \{q_i\} = \begin{Bmatrix} w_i \\ \theta_i \end{Bmatrix}\ (i=1,2) \tag{3.19}$$

The effect of transverse shear deformation is included in the new element for accurate analysis of moderately deep beams. For a slender beam, when its thickness approaches zero, λ_e in Eq. (3.10d) approaches zero as well, leading to μ_e in Eq. (3.10c) approaching 1, which, in turn, causes the transverse shear strain of the beam element approaches zero, thereby naturally avoiding the shear-locking problem.

3.3 Finite element equations and analysis procedures

The strain energy of the composite beam element is given by

$$\begin{aligned} U^{(e)} &= U_b^{(e)} + U_s^{(e)} \\ &= \frac{1}{2}\left\{q^{(e)}\right\}^T \int_0^L [B_b]^T D_{bb}[B_b]dx\left\{q^{(e)}\right\} + \frac{1}{2}\left\{q^{(e)}\right\}^T \int_0^L [B_s]^T D_{ss}[B_s]dx\left\{q^{(e)}\right\} \\ &= \frac{1}{2}\left\{q^{(e)}\right\}^T\left[K_b^{(e)}\right]\left\{q^{(e)}\right\} + \frac{1}{2}\left\{q^{(e)}\right\}^T\left[K_s^{(e)}\right]\left\{q^{(e)}\right\} \end{aligned} \tag{3.20}$$

where D_{bb} and D_{ss} are the bending stiffness and transverse shear stiffness, respectively, as given in Eqs (3.1), (3.2).

The governing equation when using the virtual work approach is

$$\left(\left[K_b^{(e)}\right] + \left[K_s^{(e)}\right]\right)\left\{q^{(e)}\right\} - \left\{f^{(e)}\right\} = 0 \tag{3.21}$$

where $\{f^{(e)}\}$ is the equivalent nodal load vector, and the element bending stiffness matrix and shear stiffness matrix can be obtained from

$$\left[K_b^{(e)}\right] = \int_0^L [B_b]^T D_{bb} [B_b] dx \tag{3.22a}$$

$$\left[K_s^{(e)}\right] = \int_0^L [B_s]^T D_{ss} [B_s] dx \tag{3.22b}$$

In this composite beam element, the Gaussian integration method is utilised, which allows the sampling points to be optimally spaced within the range of integration and achieve an accuracy as good as possible for a given number of sampling points named 'Gaussian points'. The weights and locations of the Gaussian points that range from one to six are given in Appendix B. To produce more accurate results, six Gaussian points are used for all composite beam elements in this book.

The flow chart of the finite element analysis procedures is shown in Fig. 3.3. The finite element code for this 1-D composite beam element is written in Fortran 90 for numerical analyses of homogenous and composite beams. The principal codes are provided in Appendix F.

3.4 Finite element analysis of homogeneous isotropic beams

In this section, the finite element and analysis procedures are employed to model several homogeneous beams. Before conducting the finite element analyses, a convergence study is carried out to determine the appropriate mesh size for the finite element analysis.

3.4.1 Convergence study

A homogeneous isotropic cantilevered beam (Fig. 3.4) is simulated herein. The beam is 0.554256 mm in height, 1.0 mm in width, and 4.0 mm in length. Its Young's modulus and Poisson's ratio are 2.6 MPa and 0.3, respectively. A tip load of 1.0 N is applied onto the beam. This cantilevered beam is modelled using various numbers of the composite beam element with equal lengths, and its cross section is divided into four layers. The maximum displacements calculated by the present model and the exact solution obtained from a higher-order beam theory [18] are compared in Table 3.1 and Fig. 3.5. It can be seen that the calculated maximum deflection converges to the exact solution with the increase in the number of elements and that even a very coarse mesh of only one element provides result with sufficient accuracy.

It should be mentioned that, as this cantilevered beam is a homogenous isotropic beam with linear material properties, the number of layers makes no difference in terms of its stiffness.

Fig. 3.3 Finite element solution procedures for linear analysis of composite beams.

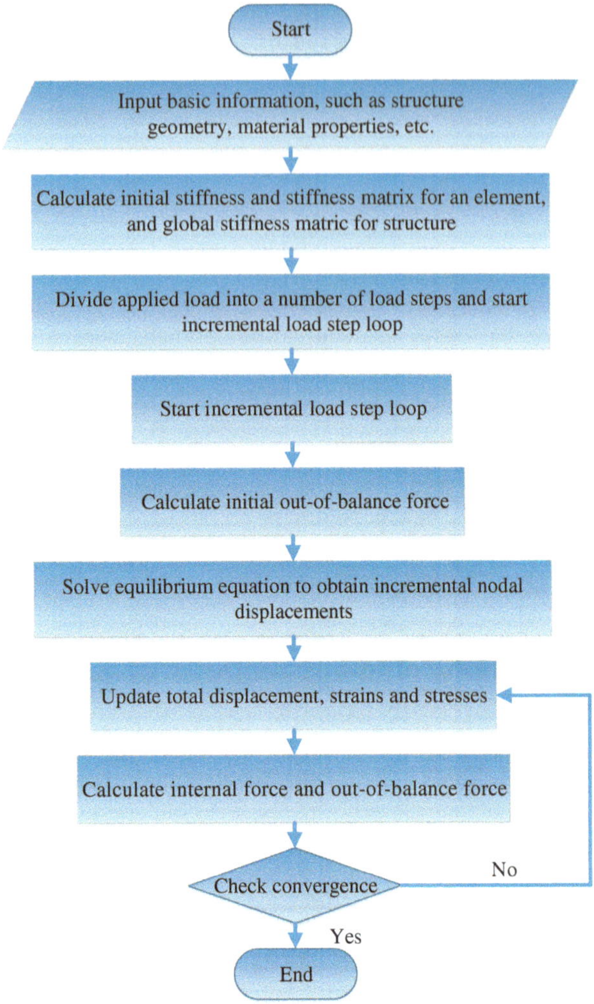

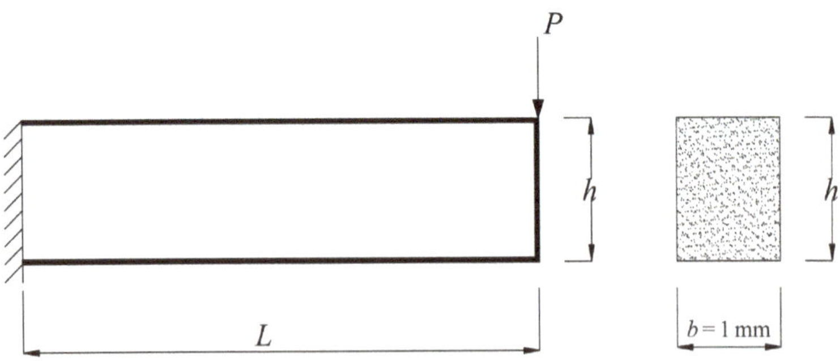

Fig. 3.4 Cantilevered beam subjected to a tip load.

Table 3.1 Convergence study of composite beam element for analysis of a homogeneous isotropic cantilevered beam.

Number of elements	Number of nodes	Maximum deflection (mm)
1	2	586.9356
4	5	586.9263
16	17	586.9059
32	33	586.9016
Exact solution [18]: 586.9		

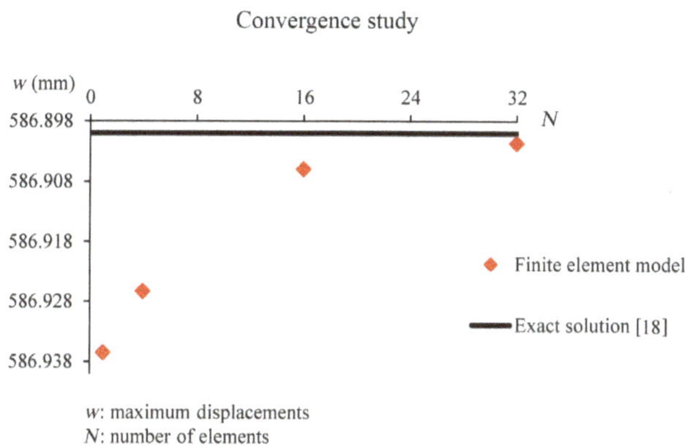

Fig. 3.5 Convergence study of a homogeneous isotropic cantilevered beam.

3.4.2 *Finite element analysis of homogeneous isotropic cantilevered beams with varying length to depth ratios*

In this section, the shear locking-free characteristic of the composite beam element is demonstrated through the numerical simulation of slender to deep beams. The cantilevered beams shown in Fig. 3.4 with various length-to-depth ratios (L/h) and constant widths of 1 mm are modelled using four of the 1-D composite beam elements with equal lengths. Each beam is subjected to a tip load P of 100 N and has a Young's modulus E and Poisson's ratio υ of 29,000 MPa and 0.3, respectively. The maximum displacements obtained from the numerical simulation are compared with those from the other finite element studies [9,18,20] and those from the Euler and Timoshenko beam theories [20] in Table 3.2.

It can be seen in Table 3.2 that the maximum displacements obtained from the present model and those from the Timoshenko's theory are exactly the same, with the former also being very close to other finite element analysis results [9,18,20]. Furthermore, the present finite element model yields accurate results for both slender beams and deep beams with length-to-depth ratios (L/h) ranging from 1.0 to 13.33.

It should be noted that the finite elements proposed by Heyliger and Reddy [18], Eisenberger [20], and Murthy et al. [9] are all based on the higher-order shear

Table 3.2 Maximum displacements of cantilevered beams with various ratios of L/h (mm).

L (mm)	**h (mm)**	**Euler theory [20]**	**Timoshenko theory [20]**	**Eisenberger [20]**	**Murthy et al. [9]**	**Heyliger and Reddy [18]**	**Present model**
160	12	32.695	32.838	32.838	32.838	32.823	32.838
80	12	4.0868	4.1586	4.1588	4.1588	4.1567	4.1586
40	12	0.5109	0.5467	0.5461	0.5461	0.5459	0.5467
12	12	0.0138	0.0246	0.0240	0.0240	0.0239	0.0246

deformation theory. The elements developed by Heyliger and Reddy [18] and Murthy et al. [9] are of two nodes and four degrees of freedom per node and that proposed by Eisenberger [20] of two nodes and three degrees of freedom per node. Moreover, 16 elements with 68 degrees of freedom were used in Heyliger and Reddy's model [18], whereas only four elements with 10 degrees of freedom are used in the current finite element model. This example shows that the new composite beam element with only two nodes and two degrees of freedom per node is not only accurate but also simple and computationally efficient.

3.4.3 Finite element analysis of two homogenous isotropic slender beams

The shear locking-free characteristic of the composite beam element can be further demonstrated through the simulation of slender beams in this example. Two homogenous isotropic slender beams are computed: one subjected to two concentrated loads and shown in Fig. 3.6 (Beam 1) and the other subjected to uniformly distributed loads and shown in Fig. 3.7 (Beam 2). Their Young's moduli E and Poisson's ratios υ and the dimensions and magnitudes of the applied loads are given in Table 3.3.

The slope at point A and deflection at point C for Beam 1 obtained from the current model using three elements of equal length are compared with those obtained from the

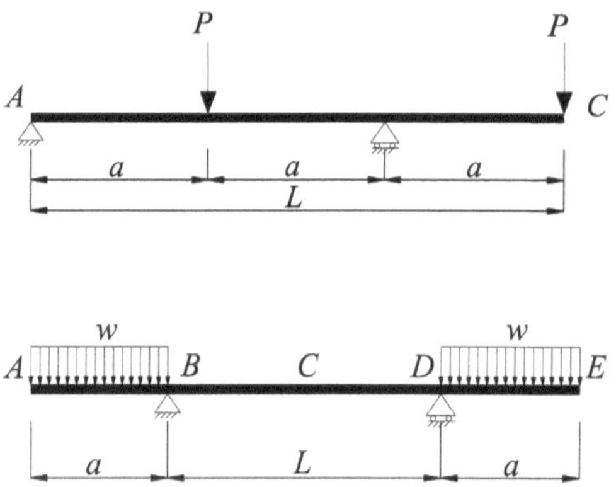

Fig. 3.6 Configuration of Beam 1 [13].

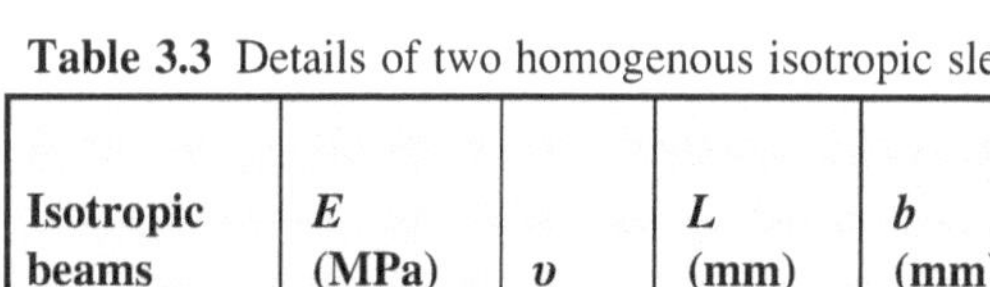

Fig. 3.7 Configuration of Beam 2 [13].

Table 3.3 Details of two homogenous isotropic slender beams.

Isotropic beams	E (MPa)	υ	L (mm)	b (mm)	h (mm)	a (mm)	P (N)	w (N/mm)
Beam 1	90000	0.3	300	10	10	100	100	–
Beam 2	90000	0.3	200	10	10	100	–	0.01

b: width of the cross section.
h: depth of the cross section.

Table 3.4 Deflection and slope of Beam 1.

Approaches	Slope at A (radian)	Deflection at C (m)
Kirchhoff [13]	1.1111×10^{-3}	1.0000×10^{-3}
Analytical solution [13]	1.1111×10^{-3}	1.0000×10^{-3}
Argyris et al. [13]	1.0938×10^{-3}	1.0052×10^{-3}
Present model	1.0966×10^{-3}	1.0077×10^{-3}

Table 3.5 Deflection and slope of Beam 2.

Approach	DOFs	Slope at E (radian)	Deflection at E (m)
Kirchhoff [13]	48	9.0122×10^{-5}	8.3946×10^{-6}
	84	8.9331×10^{-5}	8.3553×10^{-6}
Argyris et al. [13]	48	9.0123×10^{-5}	8.4124×10^{-6}
	84	8.9335×10^{-5}	8.3730×10^{-6}
Present model	10	8.9112×10^{-5}	8.3716×10^{-6}
Analytical solution [13]		8.8889×10^{-5}	8.3333×10^{-6}

analytical solution [13], those from the Kirchhoff solution [13], and those from Argyris et al. [13] in Table 3.4. The number of degrees of freedom, slope at point E, and deflection at point E for Beam 2 obtained using four composite beam elements of equal length and those from other studies [13] are listed in Table 3.5.

It can be seen that the slopes and deflections for both slender beams obtained from the numerical model using 1-D composite beam element agree very well with those calculated by classical theories [13] and Argyris et al.'s model [13]. Although the transverse shear effects have been accounted for, no shear locking occurs in both simulations. For Beam 2, when comparing the number of degrees of freedom, 48 and 84 degrees of freedom were used in the Kirchhoff model [13] and Argyris et al.'s model [13], whereas only 10 degrees of freedom are employed in the current finite element model with the 1-D composite beam element. The results obtained from the present model are even closer to the analytical solution than those from the other numerical models.

3.4.4 Finite element analysis of a homogeneous isotropic beam with I-section

The composite beam element and analysis procedures are used to analyse a simply supported homogeneous isotropic beam with an I-section, subjected to a distributed load of 5 kN/m on the top of the beam. The configuration and cross section of the beam are shown in Fig. 3.8. The Young's modulus E and Poisson's ratio υ of the I-section beam are 200 GPa and 0.3, respectively. In the finite element model, 10 composite beam elements are utilised to discretise this beam. The web and flanges of the beam are divided into 30 layers and two layers, respectively. The stresses at Point A and Point B are examined and compared in Table 3.6.

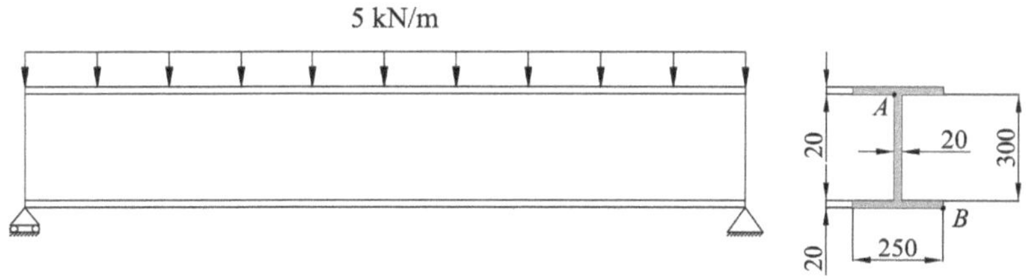

Fig. 3.8 Configuration and cross section of a homogeneous isotropic beam with I-section (unit: mm).

Table 3.6 Stresses at Point *A* and Point *B*.

Approach	Point *A* (MPa)	Point *B* (MPa)
Present model	−11.200	12.703
Analytical solution [19]	−11.201	12.694

As can be seen in Table 3.6, the stresses at both Point *A* and Point *B* obtained from the numerical model are very close to the analytical solutions [19]. It is clear that the 1-D composite beam element is versatile and suitable for beams with various cross-sectional shapes.

3.5 Finite element analysis of composite beams

In this section, the finite element and analysis procedures are employed to model two laminated composite beams.

3.5.1 Finite element analysis of a four-layered cross-ply laminated composite beam

A four-layered cross-ply laminated composite beam with the stack sequence [0°/90°/90°/0°] shown in Fig. 3.9 is modelled in this example. Its material properties are $E_1 = E_3 = 206.8$ GPa, $E_2 = 83.74$ GPa, $\upsilon_{12} = \upsilon_{13} = \upsilon_{23} = 0.12$, and $G_{12} = G_{13} = G_{23} = 48.27$ GPa, and the thicknesses of each layer are $h_1 = h_4 = 2.54$ mm and $h_2 = h_3 = 3.81$ mm. Firstly, a convergence study is carried out using different numbers of the 1-D composite beam element with equal lengths. Each layer of the laminated composite beam is subdivided into four layers, so the cross section is composed of 16 layers. As, in this example, the nonlinearity of the material property is not taken into account, the number of layers does not affect the convergence results of the finite element model. The calculated maximum deflections of the composite beam are shown in Table 3.7 and Fig. 3.10. As can be seen, the value of the maximum deflection converges with the increase in the number of elements, which verifies the finite element model for the analysis of composite beams.

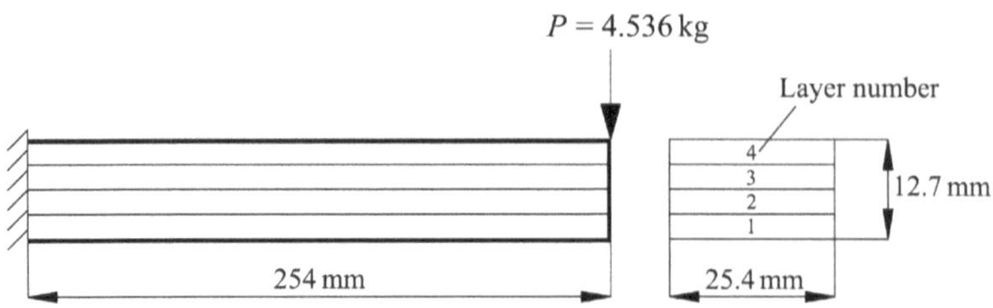

Fig. 3.9 Configuration and cross section of a cross-ply laminated composite beam [2].

Table 3.7 Convergence study of the composite beam element for analysis of a cross-ply (0°/90°/90°/0°) beam.

Number of elements	Number of nodes	*w* (mm)
4	5	0.3101457
8	9	0.3100762
16	17	0.3100436
32	33	0.3100257

Fig. 3.10 Convergence study of the composite beam element for analysis of a cross-ply laminated composite beam.

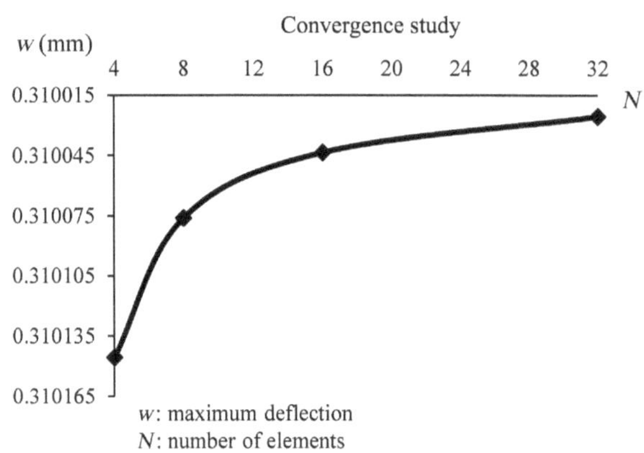

The maximum displacement obtained from the current finite element model is compared with those predicted by the models using the six-noded layered shell element PFEM [2], the eight-noded layered solid element STIF 46 [2], and the eight-noded isoparametric layered shell elements STIF 91 and 99 in ANSYS [2]. The results are listed in Table 3.8, along with the element types, the numbers of elements, and the numbers of nodes. As can be seen, even with very coarse mesh (four elements), the result from the current model is in excellent agreement with those obtained from the other finite element models, and the current model is more computationally efficient than the others as it uses much fewer elements and nodes. Moreover, compared with shell and solid elements, the 1-D beam element is much easier in modelling.

Table 3.8 Maximum deflection of a cross-ply (0°/90°/90°/0°) beam.

Element type	Number of elements	Number of nodes	*w* (mm)
PFEM [2]	16	51	0.30914
STIF 46 [2]	20	66	0.30806
STIF 91 [2]	10	53	0.30992
STIF 99 [2]	10	53	0.30992
Present element	4	5	0.31015

3.5.2 Finite element analysis of an eight-layered composite beam composed of two materials

A laminated cantilevered beam consisting of eight layers with two different materials is analysed in this section. The maximum displacement and the normal stress along its cross section are examined. The cantilevered beam is loaded with either a concentrated tip load of $Q = 200$ N (Case A) or a uniformly distributed load of $q = 100$ N/mm (Case B). The shear modulus and the Poisson's ratio for both materials are the same, 0.5×10^6 MPa and 0.25 respectively. The Young's modulus of material 1 is 30×10^6 MPa, and that for material 2 is 5×10^6 MPa. The dimensions and the cross section of this laminated cantilevered beam are shown in Fig. 3.11.

In this example, the laminated beam is modelled using six of the 1-D composite beam elements with equal length, and the cross section is discretised by 16 layers. The maximum displacements of the beam under both concentrated load and distributed load are computed and compared with the results obtained from the finite element models using other element types, including a curved beam element proposed by Surana and Nguyen [21] based on the higher-order shear deformation theory, and a layer-wise beam element with 2+*N* degrees of freedom per node (*N*: number of layers) developed by Davalos et al. [11]. It can be seen in Table 3.9 that these results are very close to each other. It is worth mentioning that 19 nodes with 646 DOFs and 17 nodes with 170 DOFs were used in the model with curved beam elements [21] and the model with layer-wise beam elements [11], respectively, whereas, in the current finite element model, only six of the composite beam elements with seven nodes and 14 DOFs are used. The current finite element model is much more cost-effective than the other two.

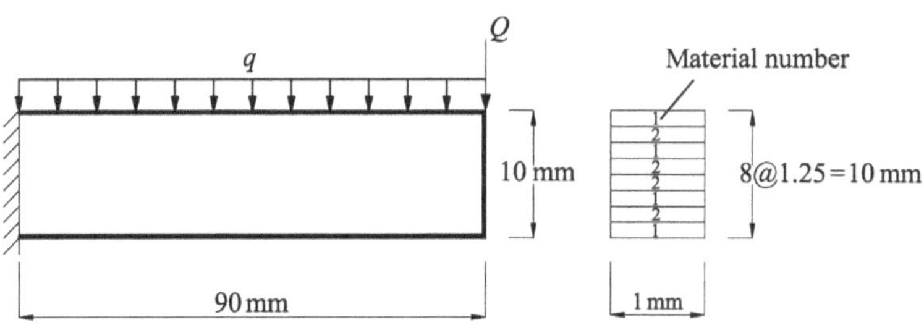

Fig. 3.11 Configuration and cross section of a laminated cantilevered beam [11].

Table 3.9 Maximum deflection of the laminated cantilevered beam (mm).

	Concentrated tip load	Uniformly distributed load
Surana and Nguyen [21]	0.03031	0.535
Davalos et al. [11]	0.03029	0.552
Present model	0.03060	0.541

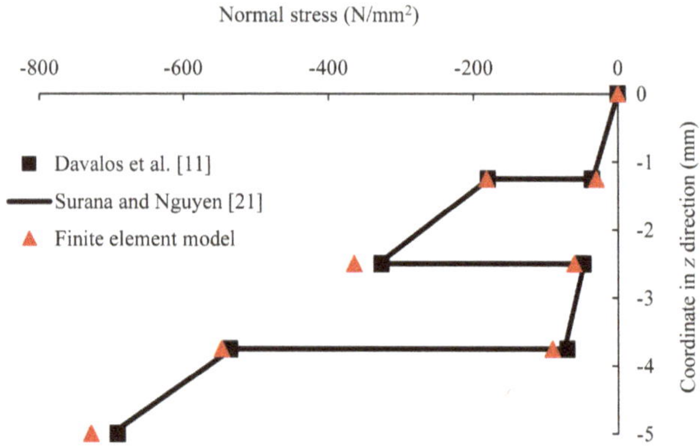

Fig. 3.12 Comparison of normal stresses at midspan of a laminated cantilevered beam (Case A).

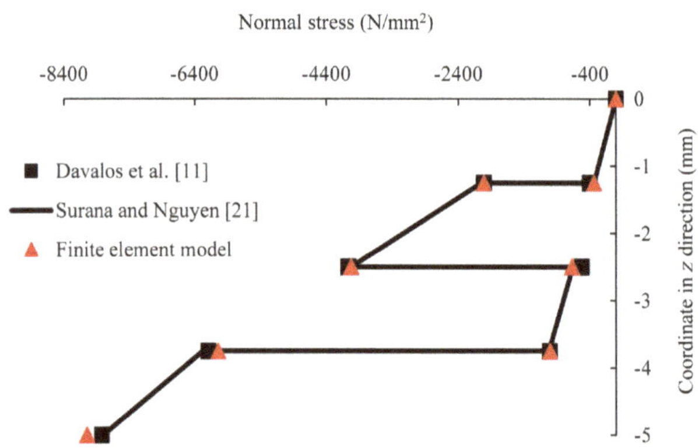

Fig. 3.13 Comparison of normal stresses at midspan of a laminated cantilevered beam (Case B).

The normal stresses at the midspan of the laminated cantilevered beam (Case A and Case B) are also computed and compared with the other two numerical models [11,21], as shown in Figs 3.12 and 3.13, respectively. It can be seen that the current results agree very well with the other two solutions.

References

[1] A.J.M. Ferreira, P.P. Camanho, A.T. Marques, A.A. Fernandes, Modelling of concrete beams reinforced with FRP re-bars, Compos. Struct. 53 (1) (2001) 107–116.
[2] H. Yu, A higher-order finite element for analysis of composite laminated structures, Compos. Struct. 28 (4) (1994) 375–383.
[3] B.S. Manjunatha, T. Kant, New theories for symmetric/unsymmetric composite and sandwich beams with C_0 finite elements, Compos. Struct. 23 (1) (1993) 61–73.
[4] M. Augusta Neto, W. Yu, S. Roy, Two finite elements for general composite beams with piezoelectric actuators and sensors, Finite Elem. Anal. Des. 45 (5) (2009) 295–304.
[5] W. Yu, D.H. Hodges, V. Volovoi, C.E.S. Cesnik, On Timoshenko-like modeling of initially curved and twisted composite beams, Int. J. Solids Struct. 39 (19) (2002) 5101–5121.
[6] F.-G. Yuan, R.E. Miller, A higher order finite element for laminated beams, Compos. Struct. 14 (1990) 125–150.
[7] P. Subramanian, Flexural analysis of laminated composite plates, Compos. Struct. 45 (1) (1999) 51–69.
[8] P. Subramanian, Flexural analysis of symmetric laminated composite beams using C1 finite element, Compos. Struct. 54 (1) (2001) 121–126.
[9] M.V.V.S. Murthy, D.R. Mahapatra, K. Badarinarayana, S. Gopalakrishnan, A refined higher order finite element for asymmetric composite beams, Compos. Struct. 67 (1) (2005) 27–35.
[10] G. Shi, K.Y. Lam, T.E. Tay, On efficient finite element modeling of composite beams and plates using higher-order theories and an accurate composite beam, Compos. Struct. 41 (1998) 159–165.
[11] J.F. Davalos, Y. Kim, E.J. Barbero, Analysis of laminated beams with a layer-wise constant shear theory, Compos. Struct. 28 (3) (1994) 241–253.
[12] P.V.R. Murthy, K.P. Rao, Finite element analysis of laminated anisotropic beams of bimodulus materials, Comput. Struct. 18 (5) (1984) 779–787.
[13] J. Argyris, L. Tenek, A. Mattsson, BEC: A 2-node fast converging shear-deformable isotropic and composite beam element based on 6 rigid-body and 6 straining modes, Comput. Method. Appl. Mech. Eng. 152 (3–4) (1998) 281–336.
[14] X. Lin, Y.X. Zhang, A novel one-dimensional two-node shear-flexible layered composite beam element, Finite Elem. Anal. Des. 47 (7) (2011) 676–682.
[15] S.B. Dong, C. Alpdogan, E. Taciroglu, Much ado about shear correction factors in Timoshenko beam theory, Int. J. Solids Struct. 47 (13) (2010) 1651–1665.
[16] W. Chen, Y.K. Cheung, Refined quadrilateral element based on Mindlin/Reissner plate theory, Int. J. Numer. Methods Eng. 47 (1-3) (2000) 605–627.
[17] W. Chen, Y.K. Cheung, Refined 9-DOF triangular Mindlin plate elements, Int. J. Numer. Methods Eng. 51 (11) (2001) 1259–1281.
[18] P.R. Heyliger, J.N. Reddy, A higher order beam finite element for bending and vibration problems, J. Sound Vib. 126 (2) (1988) 309–326.
[19] R.C. Hibbeler, Mechanics of Materials, Prentice Hall International Inc, 2004.
[20] M. Eisenberger, An exact high order beam element, Comput. Struct. 81 (2003) (2003) 147–152.
[21] K.S. Surana, S.H. Nguyen, Two-dimensional curved beam element with higher-order hierarchical transverse approximation for laminated composites, Comput. Struct. 36 (3) (1990) 499–511.

Finite element analysis of reinforced concrete beams

4

4.1 Introduction

Reinforced concrete is a principal construction material used in civil and infrastructure engineering. The history of steel-reinforced concrete structures can be traced back to 19th century. With the introduction of steel reinforcing bars in tension zone, the load bearing capacity and ductility of a structural component can be increased significantly. Lately, fibre-reinforced polymers (FRPs) have been widely used as a substitute for traditional steel reinforcements in the constructions of reinforced concrete structures, which is attributed to their superior material properties, such as high tensile strength, cost-effective fabrication, and excellent electrochemical corrosion resistance. However, the relatively low elastic modulus of FRPs may cause large deflections in structures, and their elastic brittle behaviour may result in fragile ruptures in FRP-reinforced concrete members.

During the past decades, a large number of experimental studies on the structural behaviour of FRP-reinforced concrete beams have been carried out [1–9]. Empirical relationships have also been derived to estimate the effective moment of inertia of FRP-reinforced concrete beam after cracking, and they have been built into codes, such as the ACI Building Code [10], Canadian Building Code [11], and Eurocode 2 [12]. In addition, many theoretical studies aiming to improve the empirical formulas have been conducted. For example, Qu et al. [5] investigated GFRP and steel-reinforced concrete beams based on the moment–curvature relationship for cross section to predict the load–deflection relationship of reinforced concrete beams. To obtain the effective moment of inertia and predict the deflection of a FRP-reinforced concrete beam, Bischoff [13] proposed modification factors for reducing the tension-stiffening component in Branson's equation [14]. Vogel and Svecova [15] estimated the deflections of concrete beams by establishing their cracked inertial moments whilst simultaneously considering the material properties of reinforcements and incorporating the effect of tension stiffening. Additionally, Rasheed et al. [16] presented an analytical solution for calculating the load–deflection response of a FRP-reinforced concrete beam based on integrating a rational curvature distribution along the beam using the moment area method.

Although the previously mentioned empirical equations and theoretical studies could provide direct guidance for structural design, simplified solutions would reduce accuracies in the prediction of structural behaviours, especially for large deformations of highly nonlinear materials. A performance-based analysis can provide a safe and economical design, and nonlinear finite element analysis has been demonstrated to be the most powerful and versatile tool for the accurate prediction of complicated structural behaviour.

Nonlinear Finite Element Analysis of Composite and Reinforced Concrete Beams
https://doi.org/10.1016/B978-0-12-816899-8.00004-3

Hitherto, a few finite element models have been developed for analysing the structural behaviour of FRP-reinforced concrete beams. For example, Ferreira et al. [17] modelled FRP-reinforced concrete beams using degenerated 2-D shell elements with reinforcing bars represented by an equivalent layer in the shell element. Later, this numerical model was implemented in the analysis of hybrid composite beams [18]. By using the finite element packages ANSYS and ABAQUS, Achillides and Pilakoutas [19] modelled concrete using four-node plane elements, FRP bars with two-node bar elements, and the interface between the concrete and FRP bars with nonlinear spring elements. Also, in the 3-D finite element model developed by Rafi et al. [20], 20-node solid brick elements were selected for finite element discretisation, and the reinforcing bars were modelled by bar elements embedded in the brick elements. The existing finite element models either have 2-D or 3-D characteristics in geometry or have to use bar elements to represent reinforcements, leading to expensive computational costs due to their large numbers of elements, nodes, and degrees of freedom.

In this chapter, the simple displacement-based 1-D two-node composite beam element introduced in Chapter 3 for linear analysis of composite beams is further extended for nonlinear finite element analyses of steel-/FRP-reinforced concrete beams. Similarly, Timoshenko's beam functions are employed to construct the displacement shape functions, providing a unified formulation for both slender and moderately deep beams. To represent the nonlinear material properties, the cross section of the element is divided into a number of concrete layers, and the material properties are assumed to be constant in each layer. Instead of modelling the reinforcing bars separately, the reinforcements are treated as equivalent smeared reinforcing layers and integrated into the composite beam element along with other concrete layers. The tension-stiffening effect after concrete cracking is accounted for in the material model. Additionally, geometric nonlinearity is included in the finite element model. The total Lagrangian approach and Newton–Raphson method are employed to solve the nonlinear finite element formulations.

4.2 A composite beam element for reinforced concrete beams

The 1-D layered composite beam element introduced in Chapter 3 is further extended in this chapter for nonlinear analyses of steel-/FRP-reinforced concrete beams. The composite beam element with two nodes and two degrees of freedom (transverse displacement w and rotation θ) per node and its cross section are shown in Fig. 4.1 [21], in

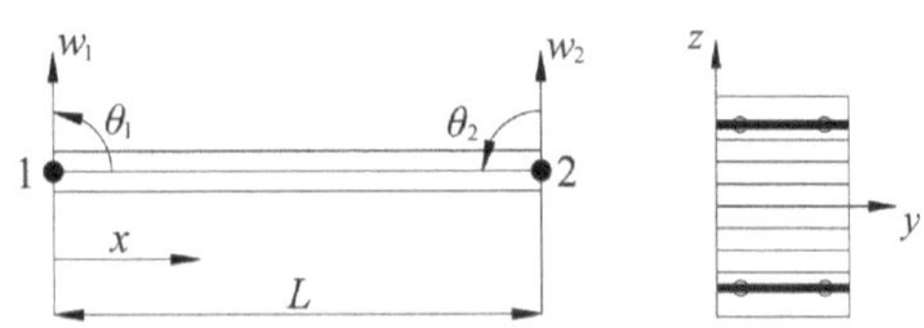

Fig. 4.1 A two-node composite beam element and its cross section [21].

which the concrete is divided into a number of concrete layers and the reinforcing bars are represented by smeared reinforcing layers with the equivalent thicknesses t given by $t = A_{bar}/b$, where A_{bar} is the cross-sectional area of the reinforcing bars and b the width of the beam element. In this element, a perfect bond between the reinforcing bars and the surrounding concrete matrix is assumed, and the local effects associated with the rebar–concrete interface, such as bond–slip and dowel action, are implicitly considered in the material model presented in Section 4.3.1.

It is assumed that each layer is in a state of plane stress and the material properties are constant throughout the thickness of each layer. The material properties of the whole section are then obtained by algebraically summing the contribution of each layer as

$$D_{bb} = b\sum\nolimits_{i=1}^{c} E_{ci} t_{ci} Y_i^2 + b\sum\nolimits_{j=1}^{s} \left(E_{sj} - E_{cj}\right) t_{sj} Y_j^2 \tag{4.1}$$

$$D_{ss} = \delta b \sum\nolimits_{i=1}^{c} G_{ci}(z_{i+1} - z_i) \tag{4.2}$$

where D_{bb} and D_{ss} are the bending stiffness and transverse shear stiffness, respectively; b the width of the beam element; c the number of concrete layers; s the number of smeared reinforcing layers; δ the constant of the nonuniformity of the shear stress, which is generally set to 5/6; E_{ci} and E_{sj} the elastic moduli of the ith concrete layer and the jth reinforcing layer, respectively; E_{cj} the elastic modulus of concrete at the level of the jth reinforcing layer; G_{ci} the shear modulus of the ith concrete layer; z_{i+1} and z_i the coordinates of the upper and lower surfaces of the ith concrete layer in the z direction, respectively; t_{ci} and t_{sj} the thicknesses of the ith concrete layer and jth reinforcing layer, respectively; and Y_i and Y_j the distances from the centroids of the ith concrete layer and jth reinforcing layer to the neutral axis, respectively.

4.2.1 Basic formulations

Assuming that the in-plane forces are zero and the in-plane displacements in the plane where the neutral axis stays are zero, the displacement field of the beam element takes the form of

$$u(x, Y) = -Y\theta(x) \tag{4.3}$$

$$w(x, Y) = w_0(x) \tag{4.4}$$

where Y is the distance from a point along the z direction to the neutral axis and w_0 the transverse displacement of the neutral axis. With the development of cracks in a reinforced concrete beam, the neutral axis is moving towards the compression zone; thus, Y in Eq. (4.3) is different from z in Eq. (3.3).

The strain–displacement relationships can be expressed as

$$\varepsilon_x = \frac{\partial u}{\partial x} = -Y\frac{d\theta}{dx} \tag{4.5}$$

$$\varepsilon_b = -\frac{d\theta}{dx} \tag{4.6}$$

$$\gamma = \frac{\partial u}{\partial Y} + \frac{\partial w}{\partial x} = \frac{\partial w}{\partial x} - \theta \tag{4.7}$$

where ε_x is the longitudinal strain of a point in the cross section at a distance Y from the neutral axis, ε_b the bending strain, γ the transverse shear strain, and $\partial w/\partial x$ the slope of the neutral axis.

The locking-free Timoshenko's beam functions [22, 23] given by Eqs (3.8), (3.9) are employed to construct the displacement shape functions, and the strain and strain matrices of the composite beam element are given in Eqs (3.16)–(3.19).

4.2.2 Nonlinear finite element formulations

To account for the geometric nonlinearity due to the possible large deformations of composite beam, the strain vector at a point along the neutral axis can be expressed as the summation of linear and nonlinear strain vectors $\{\varepsilon_0\} + \{\varepsilon_l\}$, where $\{\varepsilon_0\}$ consists of bending strain ε_b and transverse shear strain γ and ε_l is given by

$$\varepsilon_l = \frac{1}{2}\left(\frac{dw}{dx}\right)^2 \tag{4.8}$$

where

$$\left(\frac{dw}{dx}\right)^2 = [C]\left\{q^{(e)}\right\}\cdot[C]\left\{q^{(e)}\right\} \tag{4.9}$$

$$[C] = \left[-\frac{1}{L} + \mu_e\left(\frac{1}{L} - \frac{6x}{L^2} + \frac{6x^2}{L^3}\right) \quad \left(\left(\frac{1}{L} - \frac{2x}{L^2}\right) + \mu_e\left(\frac{1}{L} - \frac{6x}{L^2} + \frac{6x^2}{L^3}\right)\right)\frac{L}{2}\right.$$
$$\left.\frac{1}{L} + \mu_e\left(\frac{6x}{L^2} - \frac{1}{L} - \frac{6x^2}{L^3}\right) \quad \left(\left(\frac{2x}{L^2} - \frac{1}{L}\right) + \mu_e\left(\frac{1}{L} - \frac{6x}{L^2} + \frac{6x^2}{L^3}\right)\right)\frac{L}{2}\right] \tag{4.10}$$

Therefore,

$$\delta(\{\varepsilon_0\} + \{\varepsilon_l\}) = ([B_0] + [B_l])\delta\left\{q^{(e)}\right\} \tag{4.11}$$

in which $[B_0]$ and $[B_l]$ are the linear and nonlinear parts of the strain matrix, respectively. The total Lagrangian approach is utilised to formulate the nonlinear finite element analysis and solution procedures. The unbalanced force $\{\psi^{(e)}\}$ of the beam element can be expressed as

$$\left\{\psi^{(e)}\right\} = \left\{R^{(e)}\right\} - \int ([B_0] + [B_l])^T \{\sigma\} dx \tag{4.12}$$

where $\{R^{(e)}\}$ is the element equivalent nodal force vector due to exterior loads and $\{\sigma\}$ the internal stress vector of the element.

The tangential stiffness matrix of the element $[K_T^{(e)}]$ is given by

$$\left[K_T^{(e)}\right] = \left[K_0^{(e)}\right] + \left[K_l^{(e)}\right] + \left[K_\sigma^{(e)}\right] \tag{4.13}$$

where $[K_0^{(e)}]$ is the linear stiffness matrix, $[K_l^{(e)}]$ the large displacement matrix, and $[K_\sigma^{(e)}]$ the initial stress matrix, which are determined by the following equations:

$$\left[K_0^{(e)}\right] = \int [B_0]^T [D_t][B_0] dx \tag{4.14}$$

$$\left[K_l^{(e)}\right] = \int \left([B_0]^T [D_t][B_l] + [B_l]^T [D_t][B_0] + [B_l]^T [D_t][B_l]\right) dx \tag{4.15}$$

$$\left[K_\sigma^{(e)}\right] = \int [C]^T [N][C] dx \tag{4.16}$$

in which $[N] = [N_x \;\; 0 \;\; 0]^T$, N_x is the axial force, and $[D_t]$ is the tangential modulus matrix given by

$$[D_t] = \begin{bmatrix} b\sum_{i=1}^{C} E_{ci}t_{ci} + b\sum_{j=1}^{S} E_{sj}t_{sj} & b\sum_{i=1}^{C} E_{ci}t_{ci}Y_i + b\sum_{j=1}^{S} E_{sj}t_{sj}Y_j & 0 \\ b\sum_{i=1}^{C} E_{ci}t_{ci}Y_i + b\sum_{j=1}^{S} E_{sj}t_{sj}Y_j & D_{bb} & 0 \\ 0 & 0 & D_{ss} \end{bmatrix} \tag{4.17}$$

where D_{bb} and D_{ss} are given in Eqs (4.1), (4.2).

Thus, the iterative formulas of the Newton–Raphson method for nonlinear analysis are given as

$$\left\{\Delta q_{i+1}^{(e)}\right\} = \left[K_{T,i}^{(e)}\right]^{-1} \left\{\psi_i^{(e)}\right\} \tag{4.18a}$$

$$\left\{q_{i+1}^{(e)}\right\} = \left\{q_i^{(e)}\right\} + \left\{\Delta q_{i+1}^{(e)}\right\} \; (i = 1, 2, 3\ldots) \tag{4.18b}$$

where $[K_{T,i}^{(e)}]$ and $\{\psi_i^{(e)}\}$ are the tangential stiffness matrix and unbalanced force at the ith iteration, $\{q_i^{(e)}\}$ and $\{q_{i+1}^{(e)}\}$ the element nodal displacement vectors at the ith and $(i+1)$th iterations, and $\{\Delta q_{i+1}^{(e)}\}$ the incremental displacement vector for the $(i+1)$th iteration.

4.3 Material models

4.3.1 Material model of concrete

To simulate the nonlinear behaviour of concrete in compression, a constitutive material model [24] composed of an ascending quadratic curve and a descending linear branch as shown in Fig. 4.2 is employed. The relationship between the compressive stress σ and the compressive strain ε is given in the following equations. For simplicity, the maximum principal strain is used as yield criterion:

$$\frac{\sigma}{\sigma_{cu}}=\frac{\varepsilon}{\varepsilon_0}\left(2-\frac{\varepsilon}{\varepsilon_0}\right) \text{ for } 0\leq\varepsilon\leq\varepsilon_0 \tag{4.19a}$$

$$\frac{\sigma}{\sigma_{cu}}=1-0.15\left(\frac{\varepsilon-\varepsilon_0}{\varepsilon_{cu}-\varepsilon_0}\right) \text{ for } \varepsilon_0\leq\varepsilon\leq\varepsilon_{cu} \tag{4.19b}$$

where ε_{cu} is the ultimate compressive strain of concrete and σ_{cu} the ultimate compressive stress, given as $0.85f_c'$, in which f_c' is the concrete compressive strength, and $\varepsilon_0=1.71(f_c'/E_c)$ corresponds to the compressive strain at σ_{cu}, where E_c is the Young's modulus of concrete [25].

Concrete in tension is assumed to be isotropic and linear elastic before cracking. Once the maximum principal strain reaches the concrete ultimate tensile strain, crack occurs, and the concrete becomes orthotropic. The concrete between the cracks could carry a certain amount of tensile force normal to the cracked plane. In addition, the concrete that adheres to the reinforcing bars also contributes to the overall stiffness of the structure [26]. For reinforced concrete beams, the tension-stiffening effect should be considered when describing the interaction between concrete and reinforcing bars after concrete cracking. In the finite element model, a refined constitutive model [27] allowing for the tension-stiffening effect can be utilised, in which the influences from different reinforcements are accounted for. The material model for concrete in tension is described by Eqs (4.20a)–(4.20d) and shown in Fig. 4.3:

$$\sigma=E_c\varepsilon \text{ for } 0\leq\varepsilon\leq\varepsilon_e \tag{4.20a}$$

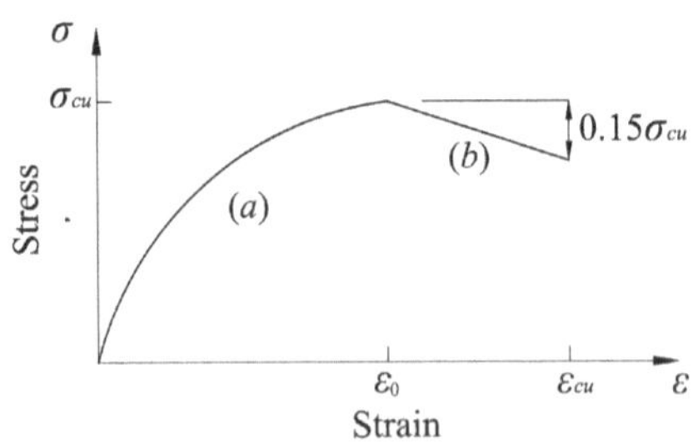

Fig. 4.2 Stress–strain relationship of concrete in compression [24].

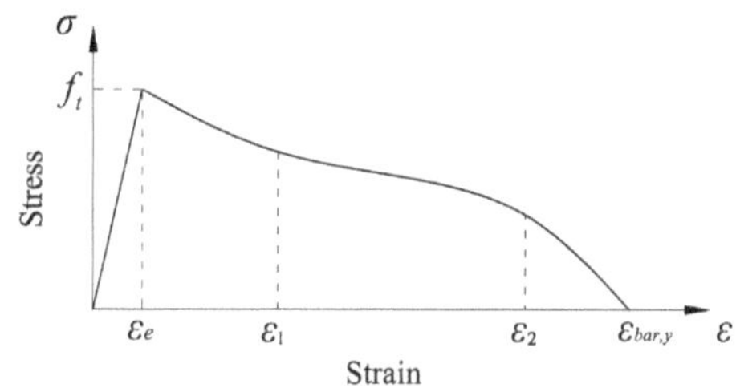

Fig. 4.3 Stress–strain relationship of concrete in tension [27].

$$\sigma = f_t\left[1-(1-\beta)\frac{\varepsilon}{\varepsilon_1}\right] \text{ for } \varepsilon_e \le \varepsilon \le \varepsilon_1 \tag{4.20b}$$

$$\sigma = \beta f_t \text{ for } \varepsilon_1 \le \varepsilon \le \varepsilon_2 \tag{4.20c}$$

$$\sigma = \beta f_t\left(\frac{\varepsilon_{bar,y}-\varepsilon}{\varepsilon_{bar,y}-\varepsilon_2}\right) \text{ for } \varepsilon_2 \le \varepsilon \le \varepsilon_{bar,y} \tag{4.20d}$$

in which σ and ε are the tensile stress and tensile strain, respectively; f_t the tensile strength of concrete; ε_e the concrete strain at f_t; and $\varepsilon_{bar,y}$ the yield strain for steel reinforcements or the ultimate strain for FRP bars.

β is the tension-stiffening factor given by Eq. (4.21) [28]:

$$\beta = \exp\left[-1100(\varepsilon-\varepsilon_e)\left(\frac{\sum E_{bar}A_{bar}}{200\sum A_{bar}}\right)\right] \tag{4.21}$$

where E_{bar} is the elastic moduli of different reinforcing bars and A_{bar} the areas of different reinforcing bars.

ε_1 and ε_2 are expressed as [27]

$$\varepsilon_1 = (1.3-\beta)f_t\left(\frac{1}{E_{bar}\rho_{eff}}+\frac{1}{E_c}\right) \tag{4.22a}$$

$$\varepsilon_2 = \varepsilon_{bar,y} - \beta f_t\left(\frac{1}{E_{bar}\rho_{eff}}+\frac{1}{E_c}\right) \tag{4.22b}$$

where ρ_{eff} is the effective reinforcement ratio, given as $\rho_{eff}=A_{bar}/A_{c,eff}$, in which $A_{c,eff}$ is the concrete area involved in stiffening and suggested as [29, 30]

$$A_{c,eff} = 2.5b(h-d_e) < \frac{b(h-d_n)}{3} \tag{4.23}$$

where h is the depth of the beam, d_e the effective depth, and d_n the depth of the neutral axis.

After cracking, shear stress is transferred across the rough surfaces of the cracked concrete through the mechanism of aggregate interlock, and the dowel action of the reinforcing bars crossing the shear cracks also contributes to the shear stiffness. To account for the effects of aggregate interlock and dowel action, the cracked shear modulus is assumed to be a function of the current tensile strain and given by [26]

$$G_c' = \begin{cases} 0.25G_c\left(1 - \dfrac{\varepsilon}{0.004}\right) & \varepsilon < 0.004 \\ 0 & \varepsilon \geq 0.004 \end{cases} \tag{4.24}$$

where G_c is the uncracked concrete shear modulus and G_c' the concrete shear modulus after cracking.

4.3.2 Material models of FRP and steel

In the numerical simulation of reinforced concrete beams, reinforcing steel bars can be considered to be elastic-perfectly plastic under both tension and compression. In the reported experimental tests [31, 32], FRP rebars have been found to have linear behaviour until they reach the point of failure without exhibiting any yielding of the material. Therefore, in the finite element model, FRP rebars are assumed to be linear elastic until the tensile stress reaches the material ultimate strength and causes brittle rupture, after which the stress within FRP immediately reduces to zero. The stress–strain relationships of steel and FRP rebars are shown in Fig 4.4A and B.

4.3.3 Calculation of neutral axis

After concrete cracking, the position of neutral axis in a cross section of a reinforced concrete beam gradually moves upwards during loading process, and the positions of neutral axis are different in different cross sections, which is attributed to the nonlinear material properties of concrete and reinforcing bars. In the finite element analysis of reinforced concrete beams, as the longitudinal strain ε_x of a point in the cross section is

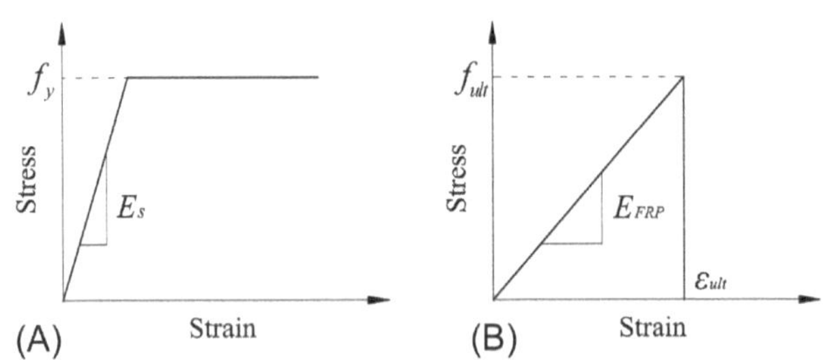

Fig. 4.4 Stress–strain relationships of (A) steel and (B) FRP.

dependent on the distance from the neutral axis, it is important to consider its varying position. The position of the neutral axis at each Gaussian point of the composite beam element can be determined using the following equation:

$$y_n = \frac{\sum_{i=1}^{c} E_{ci}\left[(0.5h - z_i)^2 - (0.5h - z_{i+1})^2\right] + \sum_{j=1}^{s}\left(E_{sj} - E_{cj}\right)\left[\left(0.5h - z_j\right)^2 - \left(0.5h - z_{j+1}\right)^2\right]}{2\left[\sum_{i=1}^{c} E_{ci} t_{ci} + \sum_{j=1}^{s}\left(E_{sj} - E_{cj}\right) t_{sj}\right]} \tag{4.25}$$

where y_n is the distance from the top of the beam to the neutral axis.

4.4 Nonlinear finite element analysis procedures

An incremental-iterative full Newton–Raphson method can be utilised to solve the nonlinear finite element equations. A displacement convergence criterion given by Eq. (4.26) is used to check the convergence of the finite element analysis:

$$\delta q_i^n \leq \lambda \Delta q_i^n \tag{4.26}$$

where δq_i^n is the iterative displacement at an iteration i in the increment n; Δq_i^n is the current incremental displacement; and λ, which is set to 0.001, is the tolerance of the displacement.

The procedures for the nonlinear finite element analyses of steel-/FRP-reinforced concrete beams are described as follows:

(a) Calculate the stiffness matrix for each beam element and then the global stiffness matrix $[K_T]$ for the entire reinforced concrete beam.
(b) Divide the applied load into a number of load increment steps.
(c) In each increment step, carry out the following iterative procedures:

- Calculate the initial out-of-balance force $\{\psi\}$, which is the subtraction of the external load and the internal force.
- Solve the finite element formulation $[K_T]\{\Delta q\} = \{\psi\}$ to obtain the incremental nodal displacements $\{\Delta q\}$ and update the total displacements.
- For each element, at the location of each Gaussian point, calculate the strain and stress increments of each layer based on the incremental nodal displacements and update the total strain and stress of each layer.
- Check the material state of each layer and update the material properties accordingly. Update the location of neutral axis for each cross section, and then, update the stiffness matrix of each element and form the global stiffness matrix.
- Calculate the internal force and out-of-balance force.
- Check the incremental displacements for convergence. If they meet the convergence criterion, go to the next load step.
- If the convergence cannot be achieved, go to the next iteration (i.e. repeat c).

The flowchart shown in Fig. 4.5 summarises the nonlinear finite element solution procedures for steel-/FRP-reinforced concrete beams, and the principal finite element codes are provided in Appendix G.

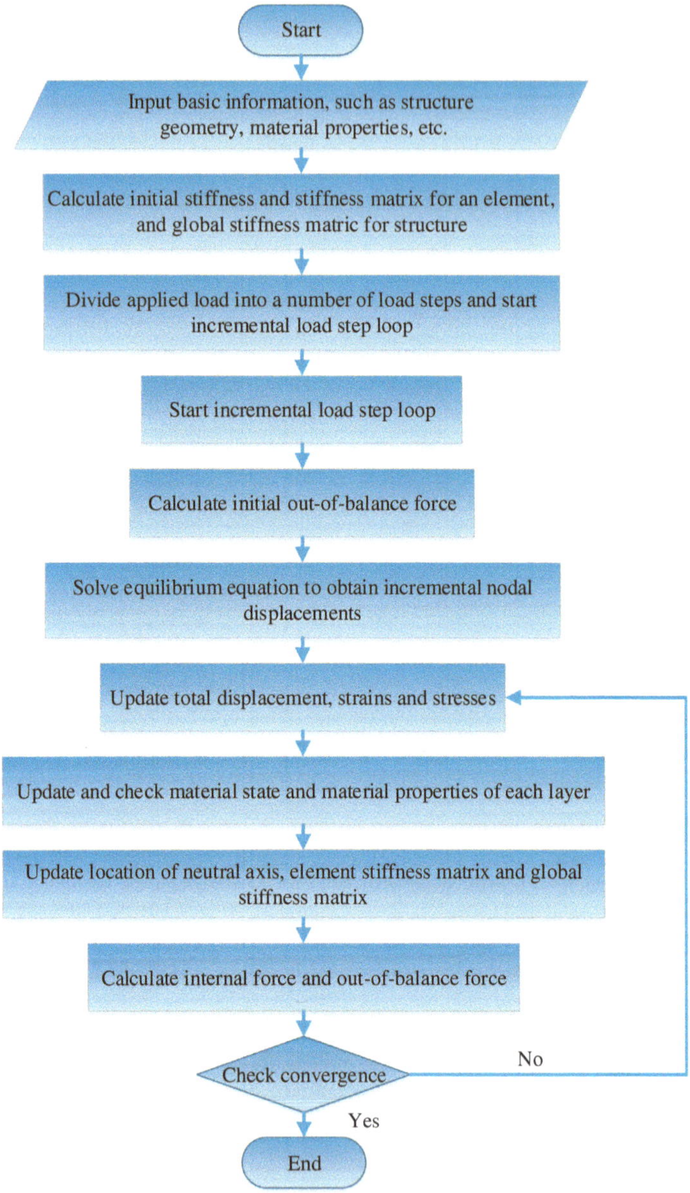

Fig. 4.5 Nonlinear finite element solution procedures for reinforced concrete beams.

4.5 Finite element analysis of reinforced concrete beams

In this section, examples of finite element analyses of reinforced concrete beams, including steel-reinforced concrete beams and FRP-reinforced concrete beams, are presented, in which the two-node composite beam elements are employed in the numerical simulations.

4.5.1 A steel-reinforced concrete beam under four-point bending load

A simply supported steel-reinforced concrete beam of 3300 mm in length, 200 mm in width, and 300 mm in depth under four-point bending load is simulated using the composite beam element. The beam is reinforced with two 19.1-mm-diameter longitudinal steel bars in tension zone and two 6-mm-diameter steel bars in compression zone, as shown in Fig. 4.6. Conventional steel stirrups spaced 100 mm apart are applied to the shear span only to prevent shear failure. The Young's moduli of both the reinforcing steel bars and stirrups are 200 GPa, and their yield strengths are 480 MPa and 300 MPa, respectively. The compressive strength of the concrete is 43 MPa, and its Young's modulus is 33 GPa.

Firstly, a convergence study is conducted. The midspan deflection of the reinforced concrete beam under an applied moment of 45 kN·m is calculated using 6, 12, 24, 30, and 48 composite beam elements of equal length, and the computed results are compared in Fig. 4.7A. It can be seen that the predicted midspan displacement converges with the increase in the number of elements and that the mesh with 12 elements is able to provide sufficiently accurate numerical estimation. It should be mentioned that the deflection obtained from the model with six elements (8.708 mm) is already very close to that obtained from the model with 48 elements (8.752 mm).

To study the effect of the number of concrete layers on the accuracy of numerical prediction, 12 composite beam elements with equal length but different numbers of concrete layers (4, 10, 20, 40, and 150) are employed to simulate the steel-reinforced concrete beam, and the calculated midspan deflections under the applied moment of 45 kN•m are compared in Fig. 4.7B. It can be seen that the result is stabilised when the number of concrete layers is 40. The discrepancies between the midspan deflections

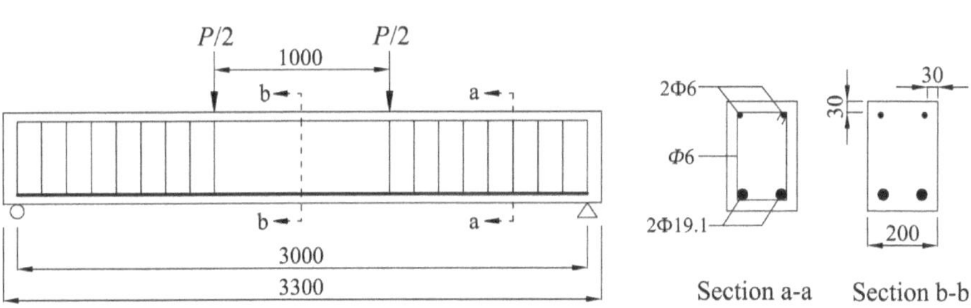

Fig. 4.6 Configurations and cross sections of a steel-reinforced concrete beam under four-point bending load (unit: mm) [9].

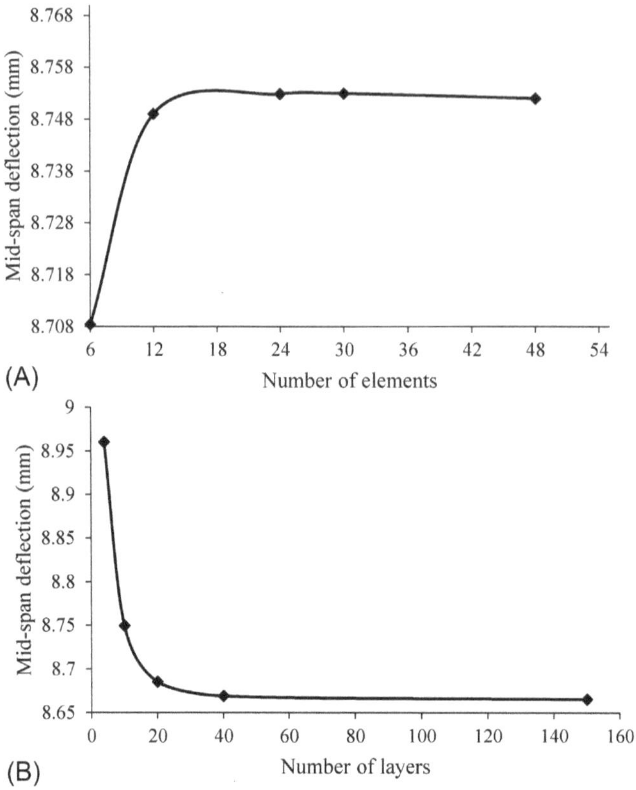

Fig. 4.7 (A) Midspan deflections of the steel-reinforced concrete beam predicted using different numbers of composite beam elements. (B) Midspan deflections of the steel-reinforced concrete beam predicted using composite beam elements with different numbers of layers.

obtained using 4, 10, 20, and 40 layers from that obtained using 150 layers are only 3%, 1%, 0.2%, and 0.05%, respectively. In this example, 40 concrete layers are selected.

The finite element mesh of the steel-reinforced concrete beam is shown in Fig. 4.8, and the predicted results are compared with those obtained from the experimental study [9], the ACI code [9], and the commercial software ANSYS in Fig. 4.9. It can be seen that the prediction from the present model agrees very well with the experimental results and ANSYS solution and is more accurate than the estimation given by the ACI code. An abrupt increase in deflection can be found in each of the solutions, which is attributed to the occurrence of concrete crack.

4.5.2 A FRP-reinforced concrete beam under four-point bending load

A FRP-reinforced concrete beam with an overall length of 2100 mm under a four-point bending load is analysed in this example. The applied loads are located 600 mm apart, giving a shear span of 600 mm. The beam is 180 mm wide and

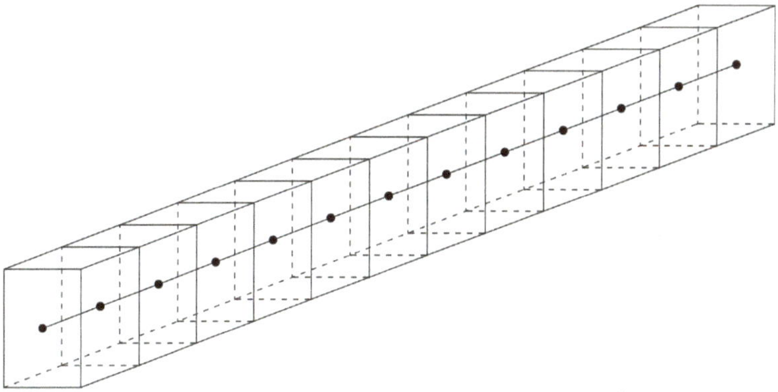

Fig. 4.8 Finite element mesh of the steel-reinforced concrete beam under four-point bending load.

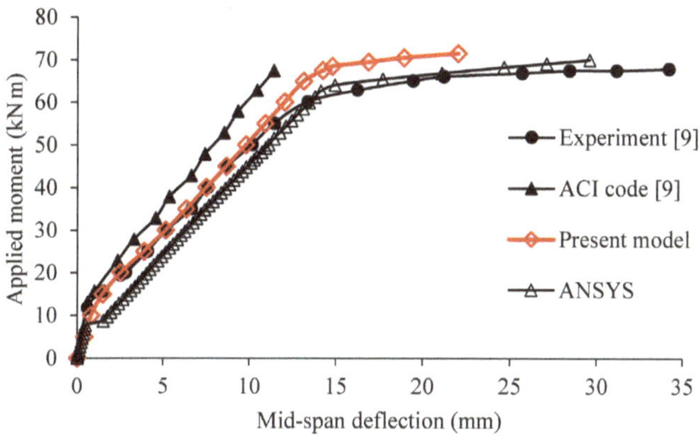

Fig. 4.9 Load–deflection relationships at midspan of the steel-reinforced concrete beam under four-point bending load.

250 mm high and reinforced with four 12.7-mm-diameter GFRP bars in its bottom region. Two 10-mm-diameter steel bars are placed in the compression zone, and 10-mm-diameter steel stirrups are used as shear reinforcements at 100-mm spacing along the shear span. Details of the GFRP-reinforced concrete beam and its cross section are shown in Fig. 4.10. The beam is modelled using six composite beam elements with equal length, and the concrete is divided into 100 layers to better capture the progress of cracking. The cubic compressive strength (f_{cu}) of concrete is 30.95 MPa, and the tensile strength and elastic modulus of the GFRP are 782 MPa and 45 GPa, respectively. The relationships between the applied load and the midspan deflection obtained from the numerical model and the experimental test [5] are compared in Fig. 4.11. It can be seen that the load–deflection curve obtained from the finite element model agrees very well with the test results attained by Qu et al. [5].

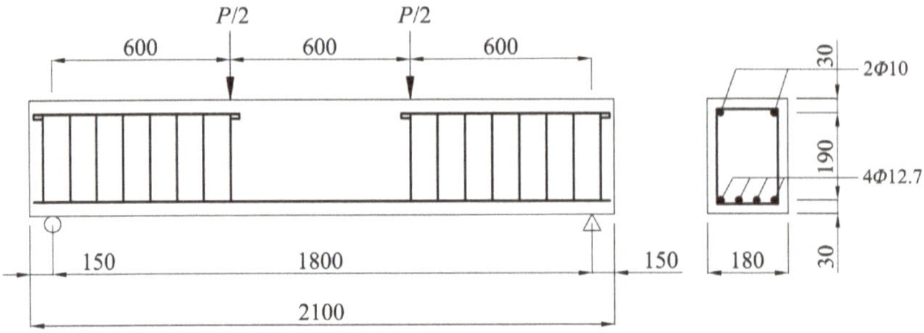

Fig. 4.10 Configuration and cross section of the GFRP-reinforced concrete beam (unit: mm) [5].

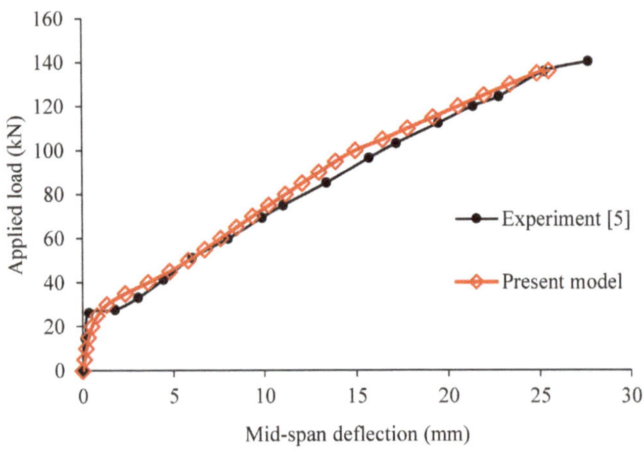

Fig. 4.11 Load–deflection relationships at midspan of the GFRP-reinforced concrete beam.

To trace the crack development progress in concrete, the maximum principal strain is checked for each iteration. Once it reaches the concrete ultimate tensile strain, crack occurs. The progress of cracking in the longitudinal section of the GFRP-reinforced concrete beam with the increase in the four-point bending load is predicted, and the results are shown in Fig 4.12A–E. It can be seen that the first cracking occurs when the applied load reaches 11 kN, then cracks develop quickly in the constant moment region with the increase in the applied load, and at a load of 15 kN the beam starts cracking out towards the supports. Significant cracking occurs when the external load is 30 kN, and cracks develop throughout the whole beam section when it comes up to the failure load.

In fact, many of these cracks, especially those developed in the interior of the concrete, are very fine at the beginning and hard to observe and measure in the experimental study, whilst the numerical simulation can effectively track the progress of crack development in concrete.

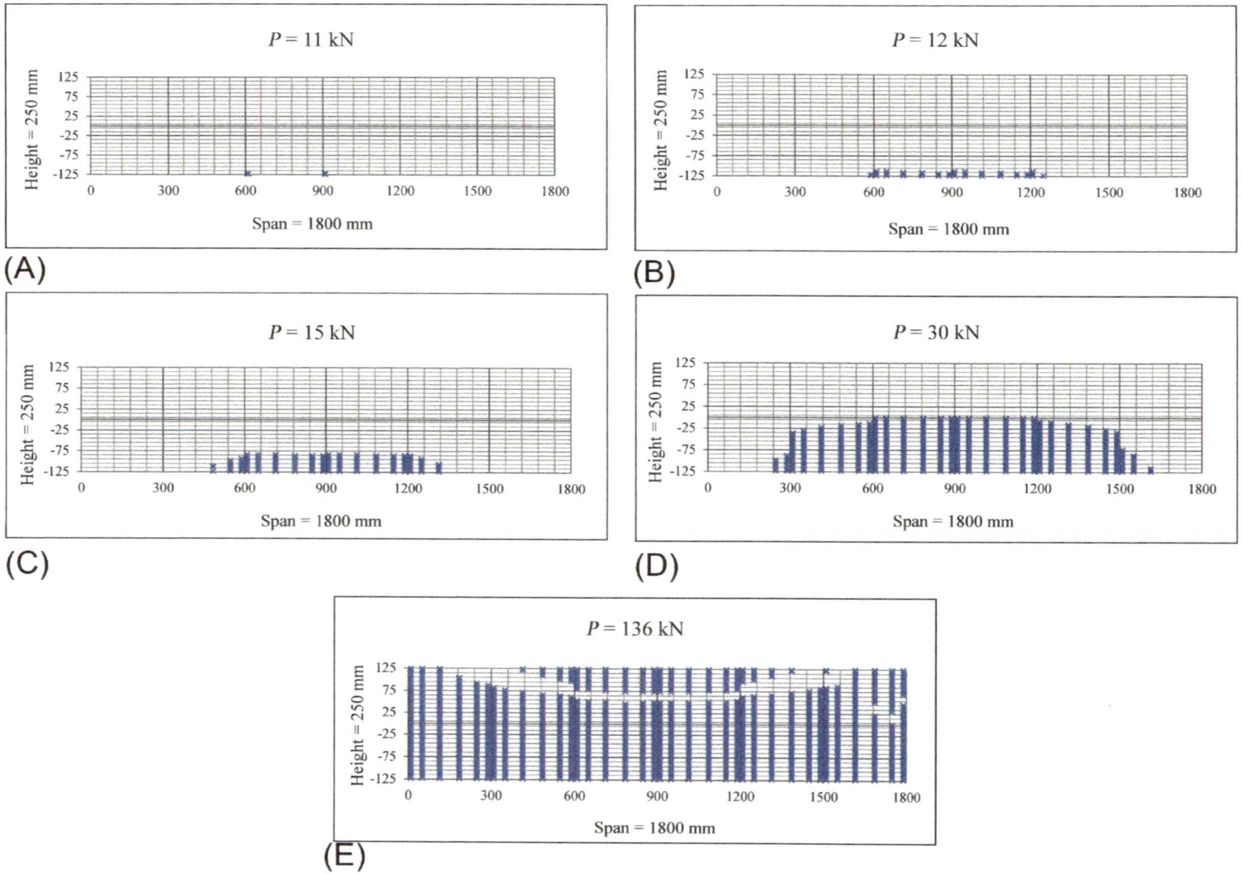

Fig. 4.12 (A) Progress of cracking at $P = 11$ kN. (B) Progress of cracking at $P = 12$ kN. (C) Progress of cracking at $P = 15$ kN. (D) Progress of cracking at $P = 30$ kN. (E) Progress of cracking at $P = 136$ kN.

References

[1] H.A. Abdalla, Evaluation of deflection in concrete members reinforced with fibre reinforced polymer (FRP) bars, Compos. Struct. 56 (2002) 63–71.
[2] H. Toutanji, Y. Deng, Deflection and crack-width prediction of concrete beams reinforced with glass FRP rods, Constr. Build. Mater. 17 (2003) (2003) 69–74.
[3] A.F. Ashour, M.N. Habeeb, Continuous concrete beams reinforced with CFRP bars, Struct. Build. 161 (2008) 349–357.
[4] C. Barris, L. Torres, A. Turon, M. Baena, A. Catalan, An experimental study of the flexural behaviour of GFRP RC beams and comparison with prediction models, Compos. Struct. 91 (2009) (2009) 286–295.
[5] W. Qu, X. Zhang, H. Huang, Flexural behavior of concrete beams reinforced with hybrid (GFRP and steel) bars, J. Compos. Constr. 13 (5) (2009) 350–359.
[6] M.A. Rashid, M.A. Mansur, P. Paramasivam, Behavior of aramid fiber-reinforced polymer reinforced high strength concrete beams under bending, J. Compos. Constr. 9 (2) (2005) 117–127.
[7] M. Pecce, G. Manfredi, E. Cosenza, Experimental response and code models of GFRP RC beams in bending, J. Compos. Constr. 4 (2000) 182–190.
[8] M.N. Habeeb, A.F. Ashour, Flexural behavior of continuous GFRP reinforced concrete beams, J. Compos. Constr. 12 (2) (2008) 115–124.
[9] B. Benmokrane, O. Chaallal, R. Masmoudi, Flexural response of concrete beams reinforced with FRP reinforcing bars, ACI Struct. J. 93 (1) (1996) 46–55.
[10] ACI Committee 440, Guide for the Design and Construction of Structural Concrete Reinforced With FRP Bars (ACI 440.1R-06), American Concrete Institute, Farmington Hills, Mich, 2006.
[11] Canadian Standards Association, Design and Construction of Building Components With Fibre-Reinforced Polymers (CAN/CSA-S806-02), 2002.
[12] European Committee for Standardization, Eurocode 2, Design of Concrete Structures. Part 1-1. General Rules for Buildings (DD ENV 1992-1-1), European Prestandard, Brussels, Belgium, 1991.
[13] P.H. Bischoff, Deflection calculation of FRP reinforced concrete beams based on modifications to the existing Branson equation, J. Compos. Constr. 11 (1) (2007) 4–14.
[14] D.E. Branson, Deformation of Concrete Structures, McGraw-Hill, New York, 1977.
[15] H. Vogel, D. Svecova, New approach for estimating the deflection of beams reinforced with FRP reinforcement, J. Compos. Constr. 12 (6) (2008) 579–587.
[16] H.A. Rasheed, R. Nayal, H. Melhem, Response prediction of concrete beams reinforced with FRP bars, Compos. Struct. 65 (2004) (2004) 193–204.
[17] A.J.M. Ferreira, P.P. Camanho, A.T. Marques, A.A. Fernandes, Modelling of concrete beams reinforced with FRP re-bars, Compos. Struct. 53 (2001) (2001) 107–116.
[18] A.J.M. Ferreira, M.C.S. Ribeiro, A.T. Marques, Analysis of hybrid beams composed of GFRP profiles and polymer concrete, Int. J. Mech. Mater. Des. 1 (2004) 143–155.
[19] Z. Achillides, K. Pilakoutas, FE modelling of bond interaction of FRP bars to concrete, Struct. Concr. 7 (1) (2006) 7–16.
[20] M.M. Rafi, A. Nadjai, F. Ali, Finite element modeling of carbon fiber-reinforced polymer reinforced concrete beams under elevated temperatures, ACI Struct. J. 105 (6) (2008) 701–710.
[21] Y.X. Zhang, X. Lin, Nonlinear finite element analyses of steel/FRP-reinforced concrete beams by using a novel composite beam element, Adv. Struct. Eng. 16 (2) (2013) 339–352.

[22] W. Chen, Y.K. Cheung, Refined quadrilateral element based on Mindlin/Reissner plate theory, Int. J. Numer. Methods Eng. 47 (1-3) (2000) 605–627.
[23] W. Chen, Y.K. Cheung, Refined 9-DOF triangular Mindlin plate elements, Int. J. Numer. Methods Eng. 51 (11) (2001) 1259–1281.
[24] C. Nitereka, K.W. Neale, Analysis of reinforced concrete beams strengthened in flexure with composite laminates, Can. J. Civ. Eng. 26 (1999) 646–654.
[25] C.A. Coronado, M.M. Lopez, Sensitivity analysis of reinforced concrete beams strengthened with FRP laminates, Cem. Concr. Compos. 28 (1) (2006) 102–114.
[26] D.R.J. Owen, J.A. Figueiras, E. Hinton, D.R.J. Owen (Eds.), Ultimate load analysis of reinforced concrete plates and shells including geometric nonlinear effects, in: Finite Element Software for Plates and Shells, Pineridge Press, Swansea, UK, 1984.
[27] A. Nour, B. Massicotte, E. Yildiz, V. Koval, Finite element modeling of concrete structures reinforced with internal and external fibre-reinforced polymers, Can. J. Civ. Eng. 34 (3) (2007) 340–354.
[28] P.H. Bischoff, R. Paixao, Tension stiffening and cracking of concrete reinforced with glass fiber reinforced polymer (GFRP) bars, Can. J. Civ. Eng. 31 (4) (2004) 579–588.
[29] Comité Euro-International du Béton, CEB-FIP model code 1990 1993. Lausanne: Thomas Telford Services Ltd.
[30] J. Zhou, J. Pan, C.K.Y. Leung, Mechanical behavior of fiber-reinforced engineered cementitious composites in uniaxial compression, J. Mater. Civ. Eng. 27 (1) (2015) 04014111.
[31] A. Nanni, T. Okamoto, M. Tanigaki, S. Osakada, Tensile properties of braided FRP rods for concrete reinforcement, Cem. Concr. Compos. 15 (3) (1993) 121–129.
[32] B. Tighiouart, B. Benmokrane, D. Gao, Investigation of bond in concrete member with fibre reinforced polymer (FRP) bars, Constr. Build. Mater. 12 (8) (1998) 453–462.

Finite element analysis of reinforced concrete beams with bond–slip

5

5.1 Introduction

In reinforced concrete structures, bond between reinforcing bars and the surrounding concrete plays an important role in transferring stress from the latter to the former and has become one of the thorny issues in the analysis of reinforced concrete structures. In many numerical simulations of reinforced concrete structures, a perfect bond between reinforcing bars and concrete has usually been assumed. Although this assumption can provide a realistic simplification of the real bonding conditions for reinforcing bars with sufficient surface preparation, when there is insufficient surface preparation, especially in FRP-reinforced concrete structures, bond–slip may take place, the effect of which on the structural behaviour should not be ignored [1–3]. In addition, with an increase in load, cracking occurs inevitably, which also results in a reduction in bond strength.

A few finite element models have been developed for analysing steel-reinforced concrete beams with bond–slip effect. In several studies, 1-D beam elements were developed based on fibre models. For example, a refined beam element proposed by Manfredi and Pecce [4] included an explicit formulation of bond–slip relationship, and each beam element was divided into subelements defined by two consecutive cracks determined by either semiempirical formulations or the spaces between stirrups. Oliveira et al. [5] presented a layered beam model based on Manfredi and Pecce's model [4], in which the bond stress–slip relationship was applied to the domain of a finite element delimited by two successive cracks. Monti and Spacone [6] developed a reinforced concrete beam finite element that combined a force-based fibre section model with a finite element model of a rebar with continuous slip. Whereas it provided the solution within a single beam element only, the determination of the force-based element state was complex. Later, a displacement-based reinforced concrete beam fibre model was presented by Spacone and Limkatanyu [7], in which the element was composed of a two-node concrete beam and several two-node bars for reinforcing steel to allow slip. It should be mentioned that the Euler–Bernoulli beam theory was applied in all these fibre model-based beam elements in which the shear effect was neglected.

In the existing 2-D and 3-D models, a separate bond element has often been used to account for the slip between reinforcing bars and the surrounding concrete. In the study reported by Kwak and Filippou [8], eight-node quadrilateral elements and 1-D truss elements were used to model the concrete and reinforcing steel, respectively, and bond-link elements were employed to consider the bond–slip effect. This method

Nonlinear Finite Element Analysis of Composite and Reinforced Concrete Beams
https://doi.org/10.1016/B978-0-12-816899-8.00005-5

is suited for cases with no significant bond–slip and associated bond damage. Khalfallah's finite element model [9] was built using eight-node serendipity plane stress elements for concrete, three-node truss elements for reinforcing bars and imperfect bond elements for bond–slip. In a 2-D model proposed by Rabczuk et al. [10], a bond element consisting of two double nodes was used to connect elements for concrete and steel. In Jendele and Cervenka's model [11], continuum elements (2-D or 3-D) were used to model concrete, truss elements with constant strain to model reinforcing bars and bond elements to model constant slip.

The existing 2-D and 3-D models for steel-reinforced concrete beams with bond–slip effect are usually complex in geometry and modelling and computationally expensive due to their large numbers of nodes and degrees of freedom. In 1-D models, inaccuracies may be introduced by using simplified methods. Therefore, a simple and effective 1-D finite element, which accounts for the effect of bond–slip, is required to enable more efficient analysis.

In this chapter, the 1-D layered composite beam element introduced in Chapter 4 is further extended for modelling reinforced concrete beams, in particular FRP-reinforced concrete beams, with bond–slip effect. In this element, the concrete is divided into a number of concrete layers and the reinforcing bars represented by equivalent smeared reinforcing layers. By using the layered method, not only can material nonlinearity be accounted for, but also the bond–slip between reinforcing bars and concrete can be simulated. In addition, both concrete and reinforcing bars can be modelled simultaneously in the integrated element, so that different types of elements are not required in numerical simulation. Apart from the degrees of freedom for transverse displacement and rotation, the composite beam element has two additional degrees of freedom to represent the axial displacements of the equivalent tensile and compressive reinforcing layers. Thus, the nodal degrees of freedom for concrete and reinforcing bars are different, allowing the reinforcing layers to slip with respect to concrete. The bond stress–slip relationships suggested in the CEB-FIP Model Code 1990 [12] and the BPE modified model [13] are used to describe the bond performances of steel and FRP reinforcing bars, respectively. As this element has only two nodes and four degrees of freedom per node, it is computationally efficient.

5.2 A composite beam element for reinforced concrete beams with bond–slip

A 1-D layered composite beam element [14], as shown in Fig. 5.1, is introduced in this chapter for nonlinear analyses of steel-/FRP-reinforced concrete beams with bond–slip effect. The layered composite beam element consists of two parts: a 1-D two-node beam element (Fig. 5.1B) for concrete and two 1-D two-node bar elements (Fig. 5.1A and C) for reinforcing bars in compression zone and tension zone, respectively.

Similar to the element introduced in Chapter 4, the cross section of the composite beam element is divided into a number of concrete layers, and the reinforcing bars are represented by equivalent smeared reinforcing layers with thicknesses t given by $t = A_{bar}/b$, as shown in Fig. 5.1E. The bending stiffness D_{bb} and the transverse shear

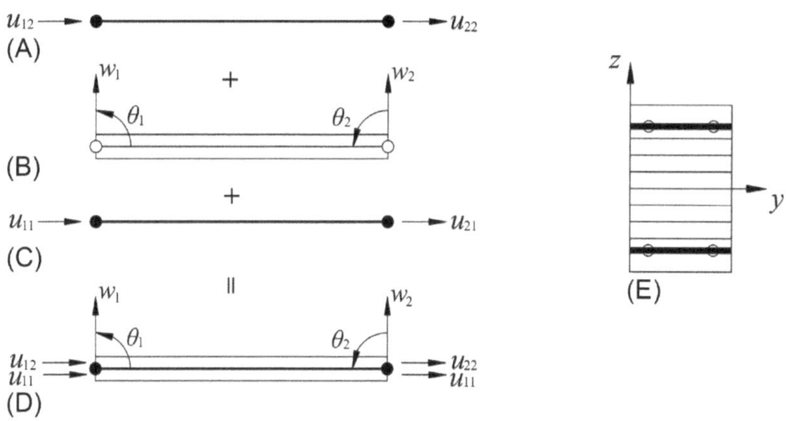

Fig. 5.1 (A) 1-D two-node bar element in compression. (B) 1-D two-node beam element. (C) 1-D two-node bar element in tension. (D) 1-D two-node beam element with bond-slip. (E) Cross section of composite beam element.
From X. Lin, Y.X. Zhang, Novel composite beam element with bond-slip for nonlinear finite element analyses of steel/FRP-reinforced concrete beams, J. Struct. Eng. 139 (12) (2013) 06013003. With permission from ASCE.

stiffness D_{ss} of the concrete beam are obtained by algebraically summing the contribution from each layer as

$$D_{bb} = b\sum_{i=1}^{c} E_{ci} t_{ci} Y_i^2 \tag{5.1}$$

$$D_{ss} = \delta b\sum_{i=1}^{c} G_{ci}(z_{i+1} - z_i) \tag{5.2}$$

To model the bond–slip behaviour between reinforcing bars and the surrounding concrete, the method proposed by Spacone and Limkatanyu [7] is employed. The equivalent smeared reinforcing layers in both tension and compression zones are represented by 1-D two-node bar elements that can slip with respect to the concrete beam. The bar elements are then merged into the layered beam element to form a new composite beam element. The nodal degrees of freedom for the reinforcing layers are separated from those for the concrete beam to permit slip, and the reinforcing layer is regarded as a system consisting of two parts: reinforcing bars and the interface between reinforcing bars and concrete. Hence, the new composite beam element has two nodes and four degrees of freedom per node (transverse displacement w, rotation θ, and axial displacement u for each of the reinforcing layers).

5.2.1 Basic formulations

The displacement field of the concrete beam takes the same form as Eqs (4.3), (4.4) as

$$u_c(x, Y) = -Y\theta(x) \tag{5.3}$$

$$w(x, Y) = w_0(x) \tag{5.4}$$

where u_c is the longitudinal displacement of a point in the cross section at a distance Y from the neutral axis.

The strain–displacement relationships of the composite concrete beam have been given in Eqs (4.5)–(4.7). When a perfect bond is assumed between concrete and reinforcing bars, the concrete and the reinforcing layers located at the same depth are of the same strain, that is,

$$\varepsilon_s = \varepsilon_c = \varepsilon_x \tag{5.5}$$

where ε_s is the strain of the reinforcing layer and ε_c the strain of the concrete layer.

With the bond–slip effect taken into account, it is assumed that the deformation of the reinforcing layer is due partly to the bar elongation and partly to the slip between the reinforcing bar and the surrounding concrete [6]. Thus, the following compatibility relation between concrete and reinforcing layer should be satisfied:

$$\varepsilon_{bond} = \varepsilon_s - \varepsilon_c \tag{5.6}$$

in which ε_{bond} is the strain due to slip. The strain–displacement relationship of the reinforcing layer is given by

$$\varepsilon_s = \frac{du_s}{dx} \tag{5.7}$$

where u_s is the axial displacement of the reinforcing layer with bond–slip.

The locking-free Timoshenko's beam functions [15, 16] are utilised to represent the transverse displacement and rotation at any point along the longitudinal direction of the composite beam element. The formulas for deflection w and rotation θ of the composite beam element with length L, width b, and height h have been given by Eqs (3.8), (3.9). The axial displacements of the bar elements representing the reinforcing layers with bond–slip are given as

For tensile reinforcement:

$$u_s = \left(1 - \frac{x}{L}\right)u_{11} + \frac{x}{L}u_{21} \tag{5.8a}$$

For compressive reinforcement:

$$u_s = \left(1 - \frac{x}{L}\right)u_{12} + \frac{x}{L}u_{22} \tag{5.8b}$$

5.2.2 Strain and strain matrix

The element bending strain, shear strain, and the strain of the reinforcing layers are expressed as

$$\varepsilon_b = [B_b]\left\{q^{(e)}\right\} \tag{5.9}$$

$$\gamma = [B_s]\left\{q^{(e)}\right\} \tag{5.10}$$

$$\varepsilon_s = [B_{bar}]\left\{q^{(e)}\right\} \tag{5.11}$$

in which

$$[B_b] = \left[\frac{6\mu_e}{L}\left(\frac{1}{L}-\frac{2x}{L^2}\right) \quad \frac{3\mu_e+1}{L} - 6\mu_e\frac{x}{L^2} \quad -\frac{6\mu_e}{L}\left(\frac{1}{L}-\frac{2x}{L^2}\right) \quad -\frac{1-3\mu_e}{L} - 6\mu_e\frac{x}{L^2} \quad 0 \;\; 0 \quad 0 \;\; 0\right] \tag{5.12a}$$

$$[B_s] = \left[-\frac{1}{L}+\frac{\mu_e}{L} \quad \frac{\mu_e}{2}-\frac{1}{2} \quad \frac{1}{L}-\frac{\mu_e}{L} \quad -\frac{1}{2}+\frac{\mu_e}{2} \quad 0 \;\; 0 \quad 0 \;\; 0\right] \tag{5.12b}$$

For tensile reinforcement:

$$[B_{bar}] = \left[0 \;\; 0 \quad 0 \;\; 0 \quad -\frac{1}{L} \;\; \frac{1}{L} \quad 0 \;\; 0\right] \tag{5.12c}$$

For compressive reinforcement:

$$[B_{bar}] = \left[0 \;\; 0 \quad 0 \;\; 0 \quad 0 \;\; 0 \quad -\frac{1}{L} \;\; \frac{1}{L}\right] \tag{5.12d}$$

where $[B_b]$ is the curvature–displacement matrix of the concrete beam, $[B_s]$ the shear strain–displacement matrix of the concrete beam, $[B_{bar}]$ the axial strain–displacement matrix of the reinforcing layers, and $\{q^{(e)}\}$ the element nodal displacement vector that can be expressed as

$$\left\{q^{(e)}\right\} = \begin{Bmatrix} q_1 \\ q_2 \end{Bmatrix} \text{ with } \{q_i\} = \begin{Bmatrix} w_i \\ \theta_i \\ u_{i1} \\ u_{i2} \end{Bmatrix} \; (i = 1, 2) \tag{5.13}$$

where u_{i1} and u_{i2} are the axial displacements for each of the reinforcing layers.

5.2.3 Nonlinear finite element formulations

As discussed in Chapter 4, to account for the geometric nonlinearity of the composite beam, the strain vector at a point along the neutral axis is given by the summation of linear and nonlinear strain vectors $\{\varepsilon_0\} + \{\varepsilon_l\}$. The nonlinear strain has been given by Eqs (4.8), (4.9) where

$$[C]=\left[-\frac{1}{L}+\mu_e\left(\frac{1}{L}-\frac{6x}{L^2}+\frac{6x^2}{L^3}\right)\quad\left(\left(\frac{1}{L}-\frac{2x}{L^2}\right)+\mu_e\left(\frac{1}{L}-\frac{6x}{L^2}+\frac{6x^2}{L^3}\right)\right)\frac{L}{2}\right.$$
$$\left.\frac{1}{L}+\mu_e\left(\frac{6x}{L^2}-\frac{1}{L}-\frac{6x^2}{L^3}\right)\quad\left(\left(\frac{2x}{L^2}-\frac{1}{L}\right)+\mu_e\left(\frac{1}{L}-\frac{6x}{L^2}+\frac{6x^2}{L^3}\right)\right)\frac{L}{2}\quad 0\ 0\ 0\ 0\right] \tag{5.14}$$

The nonlinear finite element equation for this composite beam element can be expressed as

$$\left[K_T^{(e)}\right]\left\{\Delta q^{(e)}\right\}=\left\{R^{(e)}\right\}-\int([B_0]+[B_l])^T\{\sigma\}dx-\int[B_{bar}]^T\{\sigma_{bar}\}dx \tag{5.15}$$

where $\{R^{(e)}\}$ is the element equivalent nodal loadings, $[B_0]$ and $[B_l]$ the linear and nonlinear parts of the strain matrix, $\{\sigma\}$ the internal stress vector of the concrete beam, and $\{\sigma_{bar}\}$ the internal stress vector of the reinforcements with bond–slip, which are given by the following equations:

$$\{\sigma\}=[D_t](\{\varepsilon_0\}+\{\varepsilon_l\}) \tag{5.16}$$

$$\{\sigma_{bar}\}=\sum E_{sj}A_{sj}\{\varepsilon_{sj}\}+\sum E_{bond,j}p_j\{\varepsilon_{bond,j}\} \tag{5.17}$$

where E_{sj} and $E_{bond,\,j}$ are the elastic modulus and bond modulus of the jth reinforcing layer, respectively, and A_{sj} and p_j the area and perimeter of the jth reinforcing layer, respectively. $[D_t]$ is the tangential modulus matrix of the concrete beam, given by

$$[D_t]=\begin{bmatrix} b\sum_{i=1}^{c}E_{ci}t_{ci} & b\sum_{i=1}^{c}E_{ci}t_{ci}Y_i & 0 \\ b\sum_{i=1}^{c}E_{ci}t_{ci}Y_i & D_{bb} & 0 \\ 0 & 0 & D_{ss} \end{bmatrix} \tag{5.18}$$

in which D_{bb} and D_{ss} are given in Eqs (5.1), (5.2).

The total tangential stiffness matrix of the element $[K_T^{(e)}]$ is partly contributed by the concrete beam and partly by the reinforcements with bond–slip as follows:

$$\left[K_T^{(e)}\right]=\left[K_c^{(e)}\right]+\sum\left[K_{sj}^{(e)}\right] \tag{5.19}$$

in which $[K_c^{(e)}]$ is the element stiffness matrix contributed by concrete and $[K_{sj}^{(e)}]$ the element tangential stiffness matrix contributed by the jth reinforcing layer as

$$\left[K_{sj}^{(e)}\right]=\int[B_{bar}]^T D_{bar}[B_{bar}]dx \tag{5.20}$$

in which $[B_{bar}]$ is the strain–displacement matrix of the reinforcing layers with bond–slip given by Eqs (5.12c), (5.12d).

The stiffness of the reinforcements with bond–slip D_{bar} is given by

$$D_{bar} = E_{sj}A_{sj} + E_{bond,j}p_j \tag{5.21}$$

The element stiffness matrix contributed by concrete is

$$\left[K_c^{(e)}\right] = \left[K_0^{(e)}\right] + \left[K_l^{(e)}\right] + \left[K_\sigma^{(e)}\right] \tag{5.22}$$

where $[K_0^{(e)}]$ is the element linear stiffness matrix, $[K_l^{(e)}]$ the element large displacement matrix, and $[K_\sigma^{(e)}]$ the element initial stress matrix, which have been given by Eqs (4.14)–(4.16).

5.3 Material models

The nonlinear material constitutive laws for concrete, steel, and FRP reinforcing bars discussed in Chapter 4 are also applicable to the composite beam element in this chapter. In addition, a bond stress–slip model used to predict the bond strength of the rebar–concrete interface is required.

In the full bonded range, the bond–slip model described by Eq. (5.23) [17] can be used:

$$\tau = \frac{2G_c}{d\ln\left(\overline{R}/r\right)}s \tag{5.23}$$

where τ is the shear stress at the interface between reinforcing bar and concrete, s the slip between reinforcing bar and concrete, $G_c = E_c/2(1+\upsilon)$, υ the Poisson's ratio of concrete, d the rebar diameter, and r the rebar radius, and $\ln\left(\overline{R}/r\right)$ is given as

$$\ln\left(\overline{R}/r\right) = \frac{-2\ \ln V_f - V_m\left(3 - V_f\right)}{4V_m^2} \tag{5.24}$$

in which V_f is the rebar volume fraction and $V_m = 1 - V_f$ the concrete volume fraction.

Hence, in the elastic region, the bond modulus κ [18] can be expressed as

$$\kappa = \frac{\tau}{s} = \frac{2G}{d\ \ln\left(\overline{R}/r\right)} = \frac{E}{d(1+\upsilon)\ \ln\left(\overline{R}/r\right)} \tag{5.25}$$

As the load increases, debonding takes place. A bond stress–slip model that could take into account the effect of bond–slip of reinforcing bars on the structural behaviour of reinforced concrete beams should be employed. A brief review of the currently available bond stress–slip models and the ones utilised in this book are presented in the following sections. More detailed bond stress–slip model evaluation and validation can be found in [19–21].

5.3.1 Bond stress–slip model for steel reinforcement

Up to now, a few bond stress–slip relationships have been proposed and employed in the finite element analysis of steel-reinforced concrete structures. A simple trilinear bond stress–slip model shown in Fig. 5.2A was utilised in Kwak and Kim's analytical model [23] and Khalfallah's numerical program [9]. However, this model is only applicable in the cases that do not exhibit significant bond–slip and associated bond damage. Haskett et al. [22] derived a bond stress–slip relationship (Fig. 5.2B) for deformed steel reinforcing bars, which depends on the concrete strength only.

At present, the bond–slip constitutive law developed by Eligehausen et al. [24] for deformed steel bars has been the most well-known and commonly used model in the analysis of steel-reinforced concrete structures [5, 6, 11, 25, 26]. It has an ascending curve that represents the bond mechanism of chemical adhesion, a plateau interval with constant maximum bond stress, a linear descending branch, and a final horizontal branch with a constant frictional response [27]. The model is also known as the BPE model and is adopted in the CEB-FIP Model Code 1990 [12] based on a series of parameters that depend on the amount of confinement, bond conditions, and concrete strength. Fig. 5.2C shows the bond stress–slip relation curve of this model, in which the bond stress is a function of the bond–slip s. Since then, several bond stress–slip relationships of a similar form to the Eligehausen et al.'s model [24] that consists

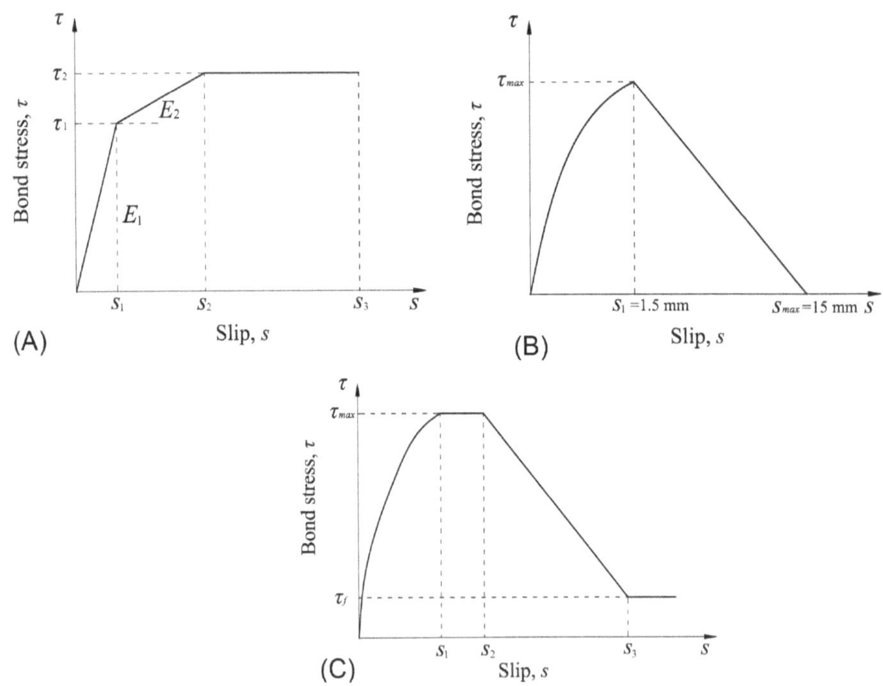

Fig. 5.2 (A) Trilinear bond stress–slip relationship of steel [9]. (B) Bond stress–slip relationship of steel proposed by Haskett et al. [22]. (C) BPE model [12].

of four stages have been developed, such as those proposed by Harajli et al. [28] and Lowes et al. [29]. In the finite element model introduced in this book, the bond stress between concrete and reinforcing steel bars under monotonic loading is calculated as a function of relative displacement s, as suggested in the CEB-FIP Model Code 1990 [12] and given in Eqs (5.26a)–(5.26d).

$$\tau = \tau_{max}(s/s_1)^{\alpha} \text{ for } 0 \leq s \leq s_1 \tag{5.26a}$$

$$\tau = \tau_{max} \text{ for } s_1 < s \leq s_2 \tag{5.26b}$$

$$\tau = \tau_{max} - (\tau_{max} - \tau_f)\left(\frac{s - s_2}{s_3 - s_2}\right) \text{ for } s_2 < s \leq s_3 \tag{5.26c}$$

$$\tau = \tau_f \text{ for } s_3 < s \tag{5.26d}$$

The parameters for the ribbed reinforcing steel in confined and unconfined concrete and the state of bond conditions for defining the bond stress–slip relationship are given in Table 5.1 [12]. Also, the parameters for defining the bond stress–slip relationship of smooth reinforcing steel are given in Table 5.2 [12].

5.3.2 Bond stress–slip model for FRP reinforcement

For FRP reinforcing bars, numerous experimental studies have been conducted to investigate the bond strength of FRP rebars in concrete and the influences of various parameters, such as fibre type, surface treatment, bar diameter, and temperature, on the bond characteristics of FRP rebars [1–3, 30, 31]. However, an extensive research effort is still needed to acquire a comprehensive bond stress–slip constitutive law for FRP rebars. A brief review of several currently available bond stress–slip models for FRP rebars is given in this section.

Table 5.1 Parameters for defining mean bond stress–slip relationship.

	Unconfined concrete		**Confined concrete**	
	Good bond conditions	**All other bond conditions**	**Good bond conditions**	**All other bond conditions**
s_1	0.6 mm	0.6 mm	1.0 mm	1.0 mm
s_2	0.6 mm	0.6 mm	3.0 mm	3.0 mm
s_3	1.0 mm	2.5 mm	Clear rib spacing	Clear rib spacing
α	0.4	0.4	0.4	0.4
τ_{max}	$2.0\sqrt{f_{ck}}$	$1.0\sqrt{f_{ck}}$	$2.5\sqrt{f_{ck}}$	$1.25\sqrt{f_{ck}}$
τ_f	$0.15\tau_{max}$	$0.15\tau_{max}$	$0.40\tau_{max}$	$0.40\tau_{max}$

Table 5.2 Parameters for defining bond stress–slip relationship of smooth bars.

	Cold drawn wire		Hot rolled bars	
	Good bond conditions	**All other bond conditions**	**Good bond conditions**	**All other bond conditions**
$s_1=s_2=s_3$	0.01 mm	0.01 mm	0.1 mm	0.1 mm
α	0.5	0.5	0.5	0.5
$\tau_{max}=\tau_f$	$0.1\sqrt{f_{ck}}$	$0.05\sqrt{f_{ck}}$	$0.3\sqrt{f_{ck}}$	$0.15\sqrt{f_{ck}}$

An analytical expression for the monotonic bond stress–slip curve given in Eq. (5.27) was proposed by Malvar [32]. To the best of the authors' knowledge, this is the first model that describes the bond stress between FRP rebars and concrete:

$$\tau=\tau_1\frac{F\left(\frac{s}{s_1}\right)+(G-1)\left(\frac{s}{s_1}\right)^2}{1+(F-2)\left(\frac{s}{s_1}\right)+G\left(\frac{s}{s_1}\right)^2} \tag{5.27}$$

where τ_1 is the peak bond stress and s_1 the slip at the peak bond stress, which are determined by the concrete tensile strength f_t and the confining axisymmetrical radial pressure σ_r given in Eqs (5.28), (5.29), respectively. F and G are the empirical constants for different bar types:

$$\tau_1/f_t=A+B\left(1-e^{-C\sigma_r/f_t}\right) \tag{5.28}$$

$$s_1=D+E\sigma_r \tag{5.29}$$

where A, B, C, D, and E are empirical constants for different bar types.

The bond–slip constitutive law (BPE model) for deformed steel bars proposed by Eligehausen et al. [24] was applied to represent the local bond stress–slip relationship of FRP rebars by Rossetti et al. [33] and Cosenza et al. [34] through calibrating the parameters based on experimental results. Later, a bond stress–slip model for FRP rebars was proposed based on the BPE model [24] by omitting its second branch with constant maximum bond stress and considering the effects of different FRP rebar surfaces. This model is known as the BPE modified model [13] (Fig. 5.3), which has been utilised in a number of studies [27, 35, 36].

In addition, the CMR model proposed by Cosenza et al. [13] for only the ascending branch can also be used as an alternative to the BPE model [24] and is defined by

$$\frac{\tau}{\tau_1}=\left[1-\exp\left(-\frac{s}{s_r}\right)\right]^{\varphi} \tag{5.30}$$

where s_r and φ are parameters based on the curve fitting of test data.

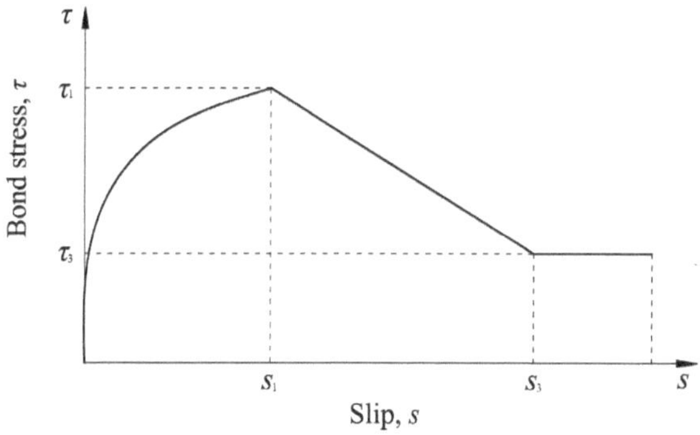

Fig. 5.3 BPE modified model [13].
From X. Lin, Y.X. Zhang, Novel composite beam element with bond-slip for nonlinear finite element analyses of steel/FRP-reinforced concrete beams, J. Struct. Eng. 139 (12) (2013) 06013003. With permission from ASCE.

Another formula for the ascending branch of the bond stress–slip relationship was proposed by Tighiouart et al. [31] based on the calibration of the parameters in the CMR model [13]. The expression of this curve is given by

$$\frac{\tau}{\tau_1} = [1 - \exp(4s)]^{0.5} \tag{5.31}$$

Although the CMR model [13] and the model proposed by Tighiouart et al. [31] are able to describe the bond stress–slip relationship between FRP rebars and concrete with good accuracy, they can only reproduce the ascending branch of the bond–slip constitutive curve, whilst Malvar's model [32] and the BPE modified model [13] are capable of simulating the entire constitutive curve. However, Malvar's model [32] has been assessed as being less reliable for modelling the ascending branch than the other models [13]. Thus, the BPE modified model [13] is employed in the finite element model in this book, and its constitutive curve is described by Eqs (5.32a)–(5.32c). It should be noted that the effect of different types of FRP reinforcing bars (i.e., CFRP, GFRP, AFRP, and BFRP) has not been taken into account in this model:

$$\frac{\tau}{\tau_1} = \left(\frac{s}{s_1}\right)^{\alpha} \text{ for } 0 \le s \le s_1 \tag{5.32a}$$

$$\frac{\tau}{\tau_1} = 1 - p\left(\frac{s}{s_1} - 1\right) \text{ for } s_1 < s \le s_3 \tag{5.32b}$$

$$\tau = \tau_3 \text{ for } s_3 < s \tag{5.32c}$$

The parameters for defining the bond stress–slip relationship of FRP rebars are given in Table 5.3.

Table 5.3 Parameters for defining bond stress–slip relationship of FRP rebars.

Reinforcement type	Ribbed	Braided	Grain-covered	Smooth
α	0.283	0.177	0.067	0.145
p	14.88	12.80	3.11	1.87
s_1	1.23 mm	2.14 mm	0.13 mm	0.26 mm
τ_1	11.61 MPa	10.20 MPa	12.05 MPa	1.19 MPa
τ_3	7.79 MPa	6.26 MPa	3.17 MPa	0.99 MPa

when $s = s_3$, we have

$$\tau = \left[1 - p\left(\frac{s_3}{s_1} - 1\right)\right]\tau_1 = \tau_3 \tag{5.33}$$

Thus,

$$s_3 = \left[\frac{1}{p}\left(1 - \frac{\tau_3}{\tau_1}\right) + 1\right]s_1 \tag{5.34}$$

To apply the bond stress–slip model to the composite beam element, the bond stress–slip response needs to be expressed in terms of the bond stress–strain curve; thus, the method used in Monti and Spacone's study [6] can be employed to obtain the bond stress–strain relationship.

In the composite beam element, the element response is the weighted sum of the response at each Gaussian point. As these Gaussian points divide the beam element into several sections, if the length of a beam element is L, the length of each section can be expressed as $L_{GP} = w_{GP}L$, where w_{GP} is the weight of the Gaussian point. Then, the slip at each Gaussian point can be expressed as

$$s = \varepsilon_{bond}L_{GP} \tag{5.35}$$

Thus, the bond strain is given by

$$\varepsilon_{bond} = \frac{s}{L_{GP}} \tag{5.36}$$

Then, the bond stress–slip relationship described by Eqs (5.26a)–(5.26d), (5.32a)–(5.32c) can be transferred into the relationship between the bond stress and bond strain.

The modulus of bond E_{bond} is given by

$$E_{bond} = \kappa L_{GP} \tag{5.37}$$

where κ is given in Eq. (5.25).

5.4 Nonlinear finite element analysis procedures

The full Newton–Raphson method presented in Chapter 4 can also be utilised to perform the nonlinear analysis in this chapter, and the displacement convergence criterion is chosen. The procedures for the nonlinear finite element analyses of steel-/FRP-reinforced concrete beams are established as follows:

(a) Calculate the stiffness matrices for both concrete beam and reinforcements with bond–slip and then the global stiffness matrix $[K_T]$ for the entire structure.
(b) Divide the applied load into a number of load increment steps.
(c) In each increment step, carry out the following iterative procedures:

- Calculate the initial out-of-balance force $\{\psi\}$, which is the subtraction of the external load and internal forces from concrete beam and reinforcements.
- Solve the finite element formulation $[K_T]\{\Delta q\} = \{\psi\}$ to obtain the incremental nodal displacements $\{\Delta q\}$ and update the total displacements.
- For each element, at the location of each Gaussian point, calculate the strain and stress increments in each layer based on the incremental nodal displacements. Also, compute the bond strain and stress of each reinforcing layer. Update the total strain and stress of each layer.
- Check the material state of each layer and update the material properties accordingly. Update the location of the neutral axis of each cross section and the bond modulus at the location of each reinforcing layer. Then, update the stiffness matrix of each element and form the global stiffness matrix.
- Calculate the internal force and out-of-balance force.
- Check the incremental displacements for convergence, and if they meet the convergence criterion, go to the next load step.
- If the convergence cannot be achieved, go to the next iteration (i.e., repeat c).

The principal finite element codes for the composite beam element with bond–slip effect are provided in Appendix H.

5.5 Finite element analysis of reinforced concrete beams with bond-slip

In this section, the finite element analysis of reinforced concrete beams with bond–slip is demonstrated in two examples: One is for steel-reinforced concrete beam, and the other is for FRP-reinforced concrete beam. The analysis results obtained from the model with bond–slip effect are also compared with those from the model with a perfect bond assumption.

5.5.1 A steel-reinforced concrete beam under four-point bending load

The simply supported steel-reinforced concrete beam modelled in Chapter 4 using the composite beam element with a perfect bond assumption is computed in this example using the element considering bond–slip effect. Details of the dimensions and cross-sectional design of the steel-reinforced concrete beam, as well as the material properties

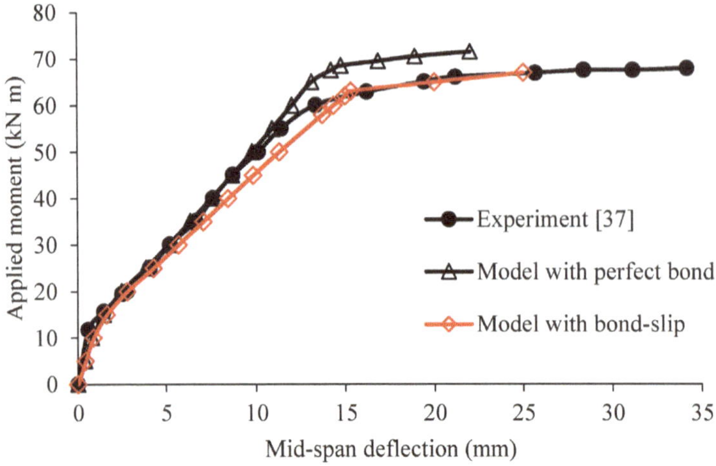

Fig. 5.4 Load–deflection relationships at midspan of steel-reinforced concrete beam under four-point bending load.

of concrete and reinforcing steel, have been given in Chapter 4 and shown in Fig. 4.6. According to the CEB-FIP Model Code 1990 [12], this concrete beam is unconfined, and as there is no detailed bond information given in the reference [37], a good bond condition is assumed.

The steel-reinforced concrete beam is modelled using 12 composite beam elements of equal length, and the concrete is divided into 150 layers to accurately simulate the nonlinear material properties of concrete. Fig. 5.4 shows the comparison of the load versus midspan deflection curves of the steel-reinforced concrete beam obtained from the finite element models with and without bond–slip effect. It can be seen that, in general, both models with and without bond–slip effect could well capture the structural behaviour of the steel-reinforced concrete beam, which is evidenced by the good agreements in between the numerical predictions and the test results [37]. With increases in the applied load, especially at the final loading stage, the model with bond–slip effect shows better capacity to describe the behaviour of the steel-reinforced concrete beam. When the load increases, the numbers and widths of cracks in concrete develop, which may lead to reduction in bond strength, especially at the end of the loading stage; when cracks develop wider and deeper, slip between concrete and reinforcements may occur. Therefore, the model with a perfect bond assumption is limited in its capability to accurately predict the structural behaviour in the full loading process, whereas the model with bond–slip effect is able to model steel-reinforced concrete beams from the state of serviceability till ultimate failure.

5.5.2 A FRP-reinforced concrete beam under four-point bending load

A CFRP-reinforced concrete beam is analysed in this example using both composite beam elements with and without bond–slip effect. The CFRP-reinforced concrete beam has an overall length of 2000 mm and a cross section of 120 mm × 200 mm.

Two 9.5-mm-diameter CFRP bars are placed in the tension zone and two 8-mm-diameter steel bars in the compression area. A 20-mm concrete cover is used all around the beam. The beam is simply supported with a span length of 1750 mm under a four-point bending load. Details of the cross section and loading setup are shown in Fig. 5.5. The compressive strength of the concrete is 41.71 MPa; the ultimate strength and elastic modulus of the CFRP rebars are 1676 MPa and 135.9 GPa, respectively; and the yield strength and elastic modulus of the steel rebars are 566 MPa and 194 GPa, respectively. The surfaces of the CFRP rebars are textured to increase their bond with concrete. Eight composite beam elements with 100 concrete layers are used to discretise the reinforced concrete beam. As the parameter for FRP with textured surface is not available in the FRP bond stress–slip model, a grain-covered surface is assumed. The comparison of the load–deflection relationships at the midspan obtained from the model with bond–slip effect, the model with a perfect bond assumption, and the experimental test [38] is shown in Fig. 5.6. Very good agreement can be observed between the predicted results obtained from the model with bond–slip effect and the test data [38]. Those from the model with a prefect bond assumption also agree well with the experimental results, except at the final loading stage, which may be attributed to some slips occurring at this stage.

In addition to the global structural behaviour, the composite beam element can also calculate the local strain on the reinforcing bars, which can be used to monitor the bond behaviour between FRP rebars and the surrounding concrete. Fig. 5.7 shows the relationships between the applied load and the tensile strain distributions on the surface of the CFRP rebar at the midspan and 275 mm from the support in the shear span of the CFRP-reinforced concrete beam. It should be noted that the locations, from which the load–strain curves are obtained from the numerical model, are the Gaussian points closest to the measured locations in the test. It can be seen that the predicted results from the finite element model agree very well with those from the test data [39]. The strain in the CFRP rebars is almost linear up to failure of the beam, which

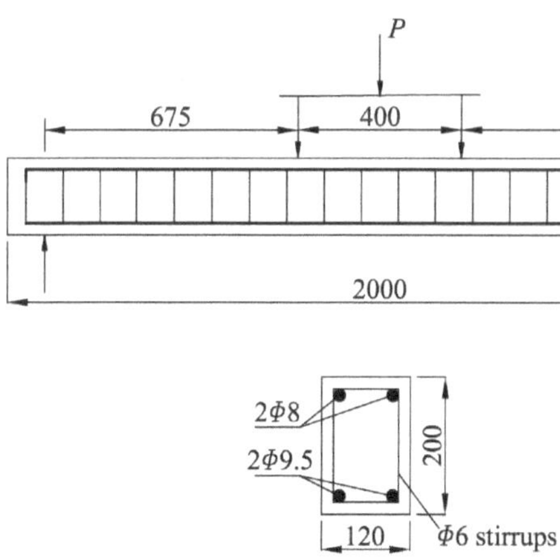

Fig. 5.5 Details of cross section and load setup of CFRP-reinforced concrete beam (unit: mm) [38].

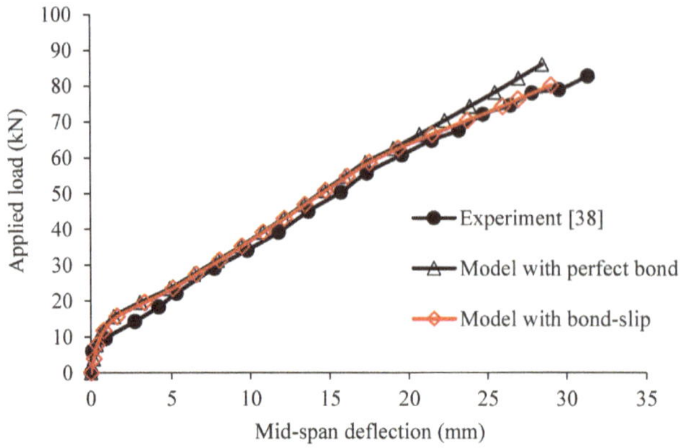

Fig. 5.6 Load–deflection relationships at midspan of CFRP-reinforced concrete beam. From X. Lin, Y.X. Zhang, Novel composite beam element with bond-slip for nonlinear finite element analyses of steel/FRP-reinforced concrete beams, J. Struct. Eng. 139 (12) (2013) 06013003. With permission from ASCE.

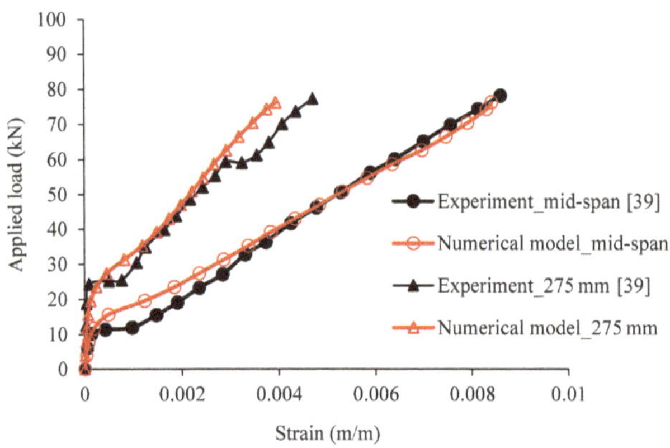

Fig. 5.7 Rebar strain distributions at midspan and 275 mm from support of CFRP-reinforced concrete beam.

is very similar to its load–deflection relationship. This phenomenon indicates a good transfer of load from concrete to the reinforcing bars, and almost no slip occurs, except at the final loading stage.

References

[1] M.A. Aiello, M. Leone, M. Pecce, Bond performances of FRP rebars-reinforced concrete, J. Mater. Civ. Eng. 19 (3) (2007) 205–213.

[2] R. Okelo, R.L. Yuan, Bond strength of fiber reinforced polymer rebars in normal strength concrete, J. Compos. Constr. 9 (3) (2005) 203–213.

[3] L.J. Malvar, J.V. Cox, K.B. Cochran, Bond between carbon fiber reinforced polymer bars and concrete. I. Experimental study, J. Compos. Constr. 7 (2) (2001) 154–163.
[4] G. Manfredi, M. Pecce, A refined R.C. beam element including bond–slip relationship for the analysis of continuous beams, Comput. Struct. 69 (1) (1998) 53–62.
[5] R.S. Oliveira, M.A. Ramalho, M.R.S. Corrêa, A layered finite element for reinforced concrete beams with bond–slip effects, Cem. Concr. Compos. 30 (3) (2008) 245–252.
[6] G. Monti, E. Spacone, Reinforced concrete fiber beam element with bond-slip, J. Struct. Eng. 126 (6) (2000) 654–661.
[7] E. Spacone, S. Limkatanyu, Responses of reinforced concrete members including bond-slip effects, ACI Struct. J. 97 (6) (2000) 831–839.
[8] H.G. Kwak, F.C. Filippou, Nonlinear FE analysis of R/C structures under monotonic loads, Comput. Struct. 65 (1) (1997) 1–16.
[9] S. Khalfallah, Tension stiffening bond modelling of cracked flexural reinforced concrete beams, J. Civ. Eng. Manag. 14 (2) (2008) 131–137.
[10] T. Rabczuk, J. Akkermann, J. Eibl, A numerical model for reinforced concrete structures, Int. J. Solids Struct. 42 (2005) 1327–1354.
[11] L. Jendele, J. Cervenka, Finite element modelling of reinforcement with bond, Comput. Struct. 84 (28) (2006) 1780–1791.
[12] Comité Euro-International du Béton, CEB-FIP Model Code 1990, Thomas Telford Services Ltd, Lausanne, 1993.
[13] E. Cosenza, G. Manfredi, R. Realfonzo, Behavior and modeling of bond of FRP rebars to concrete, J. Compos. Constr. 1 (2) (1997) 40–51.
[14] X. Lin, Y.X. Zhang, Novel composite beam element with bond-slip for nonlinear finite-element analyses of steel/FRP-reinforced concrete beams, J. Struct. Eng. 139 (12) (2013) 06013003.
[15] W. Chen, Y.K. Cheung, Refined quadrilateral element based on Mindlin/Reissner plate theory, Int. J. Numer. Methods Eng. 47 (1-3) (2000) 605–627.
[16] W. Chen, Y.K. Cheung, Refined 9-DOF triangular Mindlin plate elements, Int. J. Numer. Methods Eng. 51 (11) (2001) 1259–1281.
[17] J. Kullaa, Dimensional analysis of bond modulus in fiber pullout, J. Struct. Eng. 122 (7) (1996) 783–787.
[18] A.E. Naaman, G.G. Namur, J.M. Alwan, H.S. Najm, Fiber pullout and bond slip. I. Analytical study, J. Struct. Eng. 117 (9) (1991) 2769–2790.
[19] X. Lin, Y.X. Zhang, Evaluation of bond stress-slip models for FRP reinforcing bars in concrete, Compos. Struct. 107 (2014) 131–141.
[20] X. Lin, Y.X. Zhang, Bond–slip behaviour of FRP-reinforced concrete beams, Constr. Build. Mater. 44 (2013) 110–117.
[21] Y.X. Zhang, X. Lin, Toughening mechanisms for fiber-reinforced polymer-reinforced concrete beams, in: Q. Qin, J. Ye (Eds.), Toughening Mechanisms in Composite Materials, Woodhead Publishing, 2015, pp. 373–392 (Chapter 14).
[22] M. Haskett, D.J. Oehlers, M.S.M. Ali, Local and global bond characteristics of steel reinforcing bars, Eng. Struct. 30 (2) (2008) 376–383.
[23] H.G. Kwak, S.P. Kim, Bond–slip behavior under monotonic uniaxial loads, Eng. Struct. 23 (3) (2001) 298–309.
[24] R. Eligehausen, E.P. Popov, V.V. Bertero, Local bond stress–slip relationships of deformed bars under generalized excitations: experimental results and analytical model, in: EERC Report, Earthquake Engineering Research Center, University of California, Berkeley, 1983.

[25] M.R. Salari, E. Spacone, Finite element formulations of one-dimensional elements with bond-slip, Eng. Struct. 23 (2001) 815–826.

[26] K. Lundgren, P. Kettil, K.Z. Hanjari, H. Schlune, A.S.S. Roman, Analytical model for the bond-slip behaviour of corroded ribbed reinforcement, Struct. Infrastruct. Eng. 8 (2) (2012) 157–169.

[27] R.J. Gravina, S.T. Smith, Flexural behaviour of indeterminate concrete beams reinforced with FRP bars, Eng. Struct. 30 (9) (2008) 2370–2380.

[28] M.H. Harajli, M. Hout, W. Jalkh, Local bond stress-slip behavior of reinforcing bars embedded in plain and fiber concrete, ACI Mater. J. 92 (4) (1995) 343–353.

[29] L.N. Lowes, J.P. Moehle, S. Govindjee, Concrete-steel bond model for use in finite element modeling of reinforced concrete structures, ACI Struct. J. 101 (4) (2004) 501–511.

[30] Z. Achillides, K. Pilakoutas, Bond behavior of fiber reinforced polymer bars under direct pullout conditions, J. Compos. Constr. 8 (2) (2004) 173–181.

[31] B. Tighiouart, B. Benmokrane, D. Gao, Investigation of bond in concrete member with fibre reinforced polymer (FRP) bars, Constr. Build. Mater. 12 (8) (1998) 453–462.

[32] L.J. Malvar, Bond Stress-Slip Characteristics of FRP Rebars, California, Naval Facilities Engineering Service Center, 1994.

[33] V.A. Rossetti, D. Galeota, M.M. Giammatteo, Local bond stress-slip relationships of glass fibre reinforced plastic bars embedded in concrete, Mater. Struct. 28 (1995) 340–344.

[34] E. Cosenza, G. Manfredi, R. Realfonzo, Analytical modelling of bond between FRP reinforcing bars and concrete, in: L. Taerwe (Ed.), Nonmetallic (FRP) Reinforcement for Concrete Structures, E & FN Spon, London, 1995, pp. 164–172.

[35] M. Pecce, G. Manfredi, R. Realfonzo, E. Cosenza, Experimental and analytical evaluation of bond properties of GFRP bars, J. Mater. Civ. Eng. 13 (4) (2001) 282–290.

[36] F. Focacci, A. Nanni, C.E. Bakis, Local bond-slip relationship for FRP reinforcement in concrete, J. Compos. Constr. 4 (1) (2000) 24–31.

[37] B. Benmokrane, O. Chaallal, R. Masmoudi, Flexural response of concrete beams reinforced with FRP reinforcing bars, ACI Struct. J. 93 (1) (1996) 46–55.

[38] M.M. Rafi, A. Nadjai, F. Ali, D. Talamona, Aspects of behaviour of CFRP reinforced concrete beams in bending, Constr. Build. Mater. 22 (3) (2008) 277–285.

[39] M.M. Rafi, A. Nadjai, F. Ali, Experimental testing of concrete beams reinforced with carbon FRP bars, J. Compos. Mater. 41 (22) (2007) 2657–2673.

Finite element analysis of reinforced concrete beams at elevated temperatures

6

6.1 Introduction

As mentioned in the previous chapters, FRP reinforcing bars have been increasingly used as a substitute for traditional steel reinforcements in recent years, especially in extreme environmental conditions, which is attributed to their superior material properties. However, most of the current applications of FRP reinforcing bars are restricted to the constructions in which the temperature effect is not a primary concern, which may be due to the fact that the mechanical properties of FRP deteriorate with increases in temperature. Therefore, it is important that we understand the structural behaviour of FRP-reinforced concrete members at elevated temperatures before implementing them in engineering structures.

In the past, a few experimental studies have been carried out to investigate the effect of high temperatures on the structural performance of FRP-reinforced concrete members, for example, the thermal performance of concrete slabs reinforced with CFRP and GFRP reinforcing bars under fire conditions [1], the behaviour of CFRP-reinforced concrete beams under heating on one side with rotations at their ends restrained [2], and the performances of CFRP and hybrid steel and CFRP-reinforced concrete beams at elevated temperatures [3–5]. However, it is also well known that the experimental tests on the fire resistance of reinforced concrete structures are usually costly, time consuming, and restricted by various conditions. As such, a numerical model, which could accurately simulate the behaviour of FRP-reinforced concrete structures under a combination of thermal and mechanical loadings up to failure, is desirable. A 3-D finite element model developed by Rafi et al. [6] is one of the few numerical analyses of FRP-reinforced concrete beams under elevated temperatures. In their finite element model, 3-D solid brick elements with 20 nodes and three degrees of freedom per node were employed, and the reinforcing bars were modelled using bar elements embedded in the brick elements. Despite the effectiveness of the 3-D model, it can be computationally expensive, especially when a target to be simulated has large dimension and complex composition. To achieve effective and efficient numerical prediction, the employment of a simple and accurate finite element model is essential.

In this chapter, a 1-D composite beam element is introduced for nonlinear analyses of steel-/FRP-reinforced concrete beams at elevated temperatures. As this element has only two nodes and two degrees of freedom per node, it is simple in both geometry and formulations and more computationally efficient than most 2-D and 3-D finite element models. The Timoshenko's beam functions and the layered approach, which

Nonlinear Finite Element Analysis of Composite and Reinforced Concrete Beams
https://doi.org/10.1016/B978-0-12-816899-8.00006-7

have been employed in the composite beam elements in previous chapters, are again utilised to construct the displacement shape functions and discretise the cross section of this beam element. To take into account the effect of high temperature on the structural performance, the temperature-dependent material properties of concrete and reinforcing bars are included in the finite element model. In addition, the nonlinear finite element analysis is carried out based on the heat transfer theory, and four-node rectangular elements are used to predict the temperature distribution throughout the cross section of a reinforced concrete beam.

6.2 A composite beam element for reinforced concrete beams at elevated temperatures

Fig. 6.1 shows the 1-D composite beam element and its cross section for the analysis of reinforced concrete beams at elevated temperatures [7], which takes the same form as the one introduced in Chapter 4 with two nodes and two degrees of freedom per node. The concrete is divided into a number of concrete layers, and the reinforcing bars are treated as smeared reinforcing layers with equivalent thicknesses t given by $t = A_{bar}/b$. A perfect bond between reinforcing bars and concrete is assumed in this element, and the material properties are assumed to be constant throughout the thickness of each layer. Although FRP rebars normally show lower bond strength values compared with steel rebars, a good mechanical bond between FRP and concrete has been observed in quite a few experimental studies at room temperature [8–11], and in the fire tests conducted by Rafi and Nadjai [12, 13], the influence of temperature on the rebar slippage was found less likely to be a factor when there was a sufficient anchorage zone. In the case of bond–slip, the composite beam element can take the form as the one introduced in Chapter 5, and a temperature-dependent bond stress–slip constitutive law should be implemented in the numerical analysis. As such bond stress–slip model is currently not available for FRP, the effect of bond–slip at elevated temperatures is not considered in this book.

The Timoshenko's beam functions are again employed to construct the displacement shape functions for this composite beam element. The formulas for deflection w and rotation θ have been given in Chapter 3. The bending stiffness D_{bb} and the transverse shear stiffness D_{ss} are obtained by algebraically summing the contribution of each layer as

$$D_{bb} = b\sum_{i=1}^{c} E_{cTi}t_{ci}Y_i^2 + b\sum_{j=1}^{s}\left(E_{sTj} - E_{cTj}\right)t_{sj}Y_j^2 \tag{6.1}$$

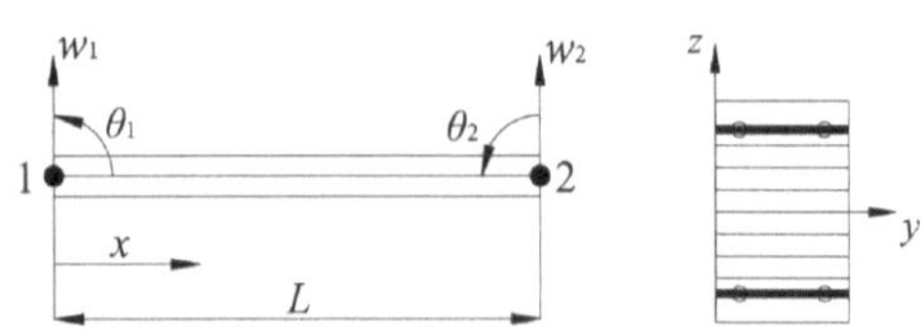

Fig. 6.1 A two-node composite beam element and its cross section [7].

$$D_{ss} = \delta b \sum_{i=1}^{C} G_{cTi}(z_{i+1} - z_i) \tag{6.2}$$

where E_{cTi} and E_{sTj} are the elastic moduli of the ith concrete layer and jth reinforcing layer at temperature T, respectively; E_{cTj} the elastic modulus of concrete at the level of the jth reinforcing layer at temperature T; and G_{cTi} the shear modulus of the ith concrete layer at temperature T, substituted by G_{cTi}^{C} after cracking.

6.2.1 Basic formulations

The displacement field and the strain–displacement relationships of the composite beam element have been given by Eqs (4.3)–(4.7). In this book, the total strain of a reinforced concrete beam under elevated temperatures is considered to be the summation of thermal expansion $\{\varepsilon_{th}\} = \{\varepsilon_{th} \ \ 0\}^T$ and mechanical strain $\{\varepsilon_{me}\}$ as

$$\{\varepsilon_{tot}\} = \{\varepsilon_{me}\} + \{\varepsilon_{th}\} \tag{6.3}$$

where $\{\varepsilon_{me}\} = \{\varepsilon \ \ \gamma\}^T$ and the thermal strains of concrete and reinforcing steel are determined according to Eurocode 2 [14] and given in Appendices B and C, respectively.

The thermal properties of FRPs vary from product to product, mainly dependent on the types of fibre and resin matrix as well as fibre volume ratio [2]. In this book, the coefficient of thermal expansion in the longitudinal direction of FRP is determined based on the values reported by Ceroni et al. [15], where T is in °C, as follows:

For GFRP rebars: $\varepsilon_{th} = 8.0 \times 10^{-6}\, T$.
For AFRP rebars: $\varepsilon_{th} = -4.0 \times 10^{-6}\, T$.
For CFRP rebars: $\varepsilon_{th} = 0$.

It is worth mentioning that the coefficient of thermal expansion of concrete is usually in the range of $7.2 \times 10^{-6}/°\text{C}$ to $10.8 \times 10^{-6}/°\text{C}$ [5]. The incompatibility in thermal expansion between concrete and reinforcing bars may lead to cracking in concrete and, thereby, FRP reinforcing bars possibly debonding from concrete. However, as mentioned earlier, the bond–slip between reinforcing bars and the surrounding concrete is not modelled in this finite element model due to the lack of a reliable bond stress–slip model for FRP rebars at high temperatures.

Therefore, the relationship between stress and mechanical strain is given as

$$\{\sigma\} = [D]\{\varepsilon_{me}\} = [D](\{\varepsilon_{tot}\} - \{\varepsilon_{th}\}) \tag{6.4}$$

where $[D]$ is the material property matrix given by

$$[D] = \begin{bmatrix} E_T & 0 \\ 0 & G_T \end{bmatrix} \tag{6.5}$$

in which E_T and G_T are the elastic modulus and shear modulus of concrete or reinforcing bars at temperature T, respectively.

The strain and strain matrix of the composite beam element have been given in Chapter 3 and Chapter 4.

6.2.2 Two-dimensional nonlinear heat transfer analysis

The temperature distribution throughout the cross section of a reinforced concrete beam can be determined using the finite element method based on the nonlinear heat transfer analysis. In this model, the cross section of a reinforced concrete beam is divided into a number of rectangular elements, and it is assumed that the reinforced concrete beam is exposed to high temperatures from three sides, as shown in Fig. 6.2. The temperature is firstly calculated at each node of each rectangular element based on its boundary conditions and the time of temperature exposure. To carry out the structural analysis using the 1-D composite beam element introduced in this book, a simplified method is used to convert the 2-D analysis into a 1-D result. In this method, the temperature at the middle point of each rectangular element is calculated based on its nodal temperatures. The temperature-dependent material properties are then calculated for each rectangular element based on its middle point temperature. Finally, the material properties of each layer in the composite beam element are assumed to be equal to the average material properties of rectangular elements in the same layer. In the present model, the temperature-dependent material thermal properties determined according to Eurocode 4 [16] are taken into account in the nonlinear heat transfer analysis.

The effect of reinforcing bars on the temperature distribution in the cross section of a reinforced concrete beam is neglected due to the small area they occupy compared with that of the concrete. In addition, in the experimental results reported in [3, 17] for CFRP-reinforced concrete beams under fire conditions, no difference was observed in the temperature distribution along the length of the beam, and the temperatures at all points on the same elevation of a longitudinal section could be considered homogeneous. Thus, in this book, it is assumed that there is no change in the temperature distribution along the length of a reinforced concrete beam.

A brief review of the finite element method for heat transfer analysis and the element and coefficients used in the current numerical model are presented herein. The governing equation for transient heat conduction is given as

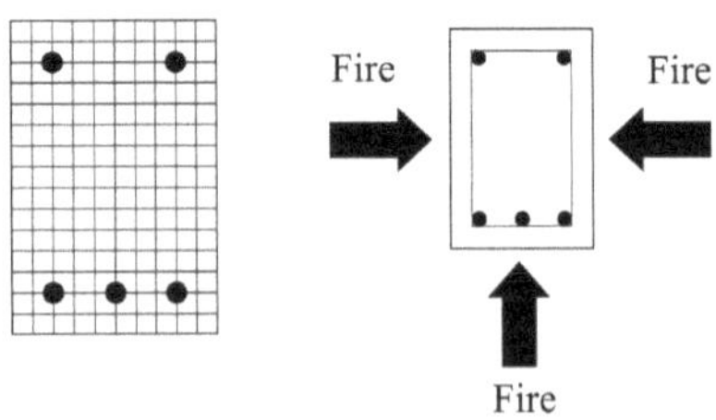

Fig. 6.2 Discretisation and heating of cross section of reinforced concrete beam in 2-D heat transfer analysis.

$$k_{cc}\left(\frac{\partial^2 T}{\partial x^2}+\frac{\partial^2 T}{\partial y^2}\right)+Q=\rho_c c_c\frac{\partial T}{\partial t} \tag{6.6}$$

where k_{cc} is the concrete thermal conductivity, Q (kW/m^3) the internal heat source, and ρ_c and c_c the concrete mass density and specific heat, respectively.

The thermal conductivity of a material is a measure of its ability to conduct heat, which depends on its composition [6]. For normal weight concrete, the thermal conductivity k_{cc} (W/m °C) suggested by Eurocode 4 [16] is utilised and given in Appendix C. The specific heat c_c (J/kg °C) and mass density ρ_c (kg/m^3) of normal weight concrete are dependent on concrete temperature and approximated by equations given in Eurocode 4 [16] (refer to Appendix C).

The boundary conditions associated with Eq. (6.6) can be expressed as

$$k_{cc}\left(\frac{\partial T}{\partial x}n_x+\frac{\partial T}{\partial y}n_y\right)+q_h=0 \tag{6.7}$$

where n_x and n_y are the components of the vector normal to the boundary in the plane of cross section, and the heat flow by convective heat transfer q_h (kW/m^2) is given by Newton's law of cooling, which is

$$q_h=h_T(T-T_\infty) \tag{6.8}$$

where h_T (W/m^2 °C) is the convection coefficient and T_∞ the fire or ambient temperature depending on the type of exposure. For a fire-exposed surface, $h_T=25$, and for a surface not exposed to fire, $h_T=9$ [18].

Four-node rectangular elements (Fig. 6.3) are used in the current numerical model for heat transfer analysis, and the temperature within each element is given by

$$T^{(e)}=[N_T]\left\{t^{(e)}\right\} \tag{6.9}$$

in which $\{t^{(e)}\}$ is the element nodal temperature vector given by

$$\left\{t^{(e)}\right\}=\{t_1\ \ t_2\ \ t_3\ \ t_4\}^T \tag{6.10}$$

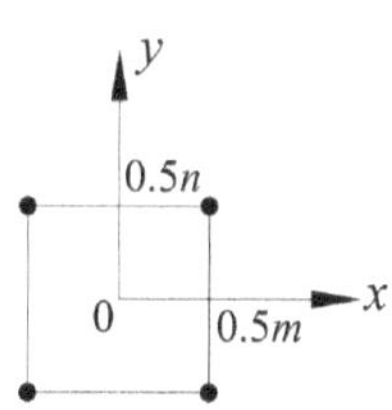

Fig. 6.3 Four-node rectangular element.

The shape function matrix $[N_T]$ can be expressed as

$$[N_T]=[N_1 \ \ N_2 \ \ N_3 \ \ N_4] \tag{6.11}$$

$$N_1=\frac{1}{mn}(0.5m-x)(0.5n-y)$$

$$N_2=\frac{1}{mn}(0.5m+x)(0.5n-y)$$

$$N_3=\frac{1}{mn}(0.5m-x)(0.5n+y)$$

$$N_4=\frac{1}{mn}(0.5m+x)(0.5n+y)$$

where m and n are the dimensions of the element in x- and y-directions, respectively.

The principal equation for the time-dependent heat transfer analysis of each element is given by

$$\left\{F_T^{(e)}\right\}=\left[K_T^{(e)}\right]\left\{t^{(e)}\right\}+\left[M^{(e)}\right]\left\{\dot{t}^{(e)}\right\} \tag{6.12}$$

where $[K_T^{(e)}]$ and $[M^{(e)}]$ are the element conduction matrix and mass matrix, respectively, $\left\{\dot{t}^{(e)}\right\}$ the temperature derivative with respect to time, and $\{F_T^{(e)}\}$ the thermal load vector. They are given by the following equations:

$$\left[K_T^{(e)}\right]=\iiint_V [B_T]^T[D_T][B_T]dV+\iint_S h_T[N_T]^T[N_T]dS \tag{6.13}$$

where

$$[B_T]=\begin{bmatrix}\frac{\partial N_1}{\partial x} & \frac{\partial N_2}{\partial x} & \frac{\partial N_3}{\partial x} & \frac{\partial N_4}{\partial x}\\ \frac{\partial N_1}{\partial y} & \frac{\partial N_2}{\partial y} & \frac{\partial N_3}{\partial y} & \frac{\partial N_4}{\partial y}\end{bmatrix} \tag{6.14}$$

The material property matrix is

$$[D_T]=\begin{bmatrix}k_{xx} & 0\\ 0 & k_{yy}\end{bmatrix} \tag{6.15}$$

in which k_{xx} and k_{yy} are the thermal conductivities in x- and y-directions, respectively.

$$\left[M^{(e)}\right] = \iiint_V c_c \rho_c [N_T]^T [N_T] dV \tag{6.16}$$

$$\left\{F_T^{(e)}\right\} = \iiint_V [N_T]^T Q dV + \iint_S q^* [N_T]^T dS + \iint_S h_T T_\infty [N_T]^T dS \tag{6.17}$$

where Q (kW/m^3) and q^* (kW/m^2) are the internal heat source and prescribed heat flux, respectively.

The global equation for the time-dependent heat transfer analysis is obtained by assembling the element principal equations into the global system, and the numerical time integration method given by Eq. (6.18) [19] is used to solve the transient heat transfer problem.

$$\left(\frac{1}{\Delta t}[M] + \eta[K_T]\right)\{t_{i+1}\} = \left(\frac{1}{\Delta t}[M] - (1-\eta)[K_T]\right)\{t_i\} + (1-\eta)\{F_{T,i}\} + \eta\{F_{T,i+1}\} \tag{6.18}$$

where i refers to the number of time steps and η is a constant between 0 and 1. To achieve an unconditionally stable solution and give less oscillatory errors [19], η is chosen to be 2/3 in the present model.

6.3 Temperature-dependent material models

The majority of construction materials would lose their strengths at high temperatures, which may severely affect the performances of engineering structures. Therefore, to accurately predict the behaviour of a reinforced concrete structure under fire conditions, the temperature-dependent material models are essential, and those employed in this book for concrete, steel, and FRP rebars are presented in this section.

6.3.1 Temperature-dependent material model of concrete

The effect of high temperatures on the mechanical properties of concrete is well documented in many references [14, 16, 20–23]. Both modulus of elasticity and compressive strength of concrete would decrease with increases in temperature. For concrete in compression, the constitutive relationships for concrete with both siliceous aggregates and calcareous aggregates suggested in Eurocode 2 [14] and Eurocode 4 [16] are employed in the current finite element model. Fig. 6.4 shows the constitutive relationship for concrete in compression at elevated temperatures.

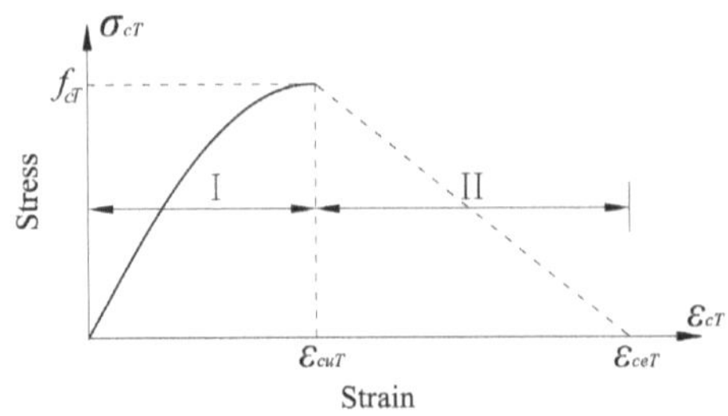

Fig. 6.4 Stress–strain relationship for concrete under compression at elevated temperatures [14, 16].

The relationship between the compressive stress σ_{cT} and strain ε_{cT} at temperature T in Range I in Fig. 6.4 is given by Eq. (6.19), and a linear descending branch is adopted in the present model for Range II:

$$\sigma_{cT} = f_{cT}\left[3\left(\frac{\varepsilon_{cT}}{\varepsilon_{cuT}}\right) \Big/ \left\{2 + \left(\frac{\varepsilon_{cT}}{\varepsilon_{cuT}}\right)^3\right\}\right] \tag{6.19}$$

where f_{cT} is the concrete compressive strength at temperature T (°C) given by Eq. (6.20), ε_{cuT} the strain corresponding to f_{cT}, and ε_{ceT} the ultimate compressive strain of the concrete at temperature T.

$$f_{cT} = k_c f_c \tag{6.20}$$

where f_c is the concrete compressive strength at ambient temperature (20°C) and k_c the reduction factor of the concrete compressive strength at temperature T. The reduction factors for concrete with calcareous aggregates and siliceous aggregates are determined according to Eurocode 2 [14] and Eurocode 4 [16], respectively, and given in Appendix C. ε_{cuT} and ε_{ceT}, which vary with temperature, are taken from Eurocode 4 [16] and listed in Appendix C as well.

Since there is insufficient data available for describing concrete constitutive relationship in tension at elevated temperatures, the concrete tensile strength f_{tT}, elastic modulus E_{cT} and shear modulus G_{cT} at temperature T (°C) are assumed to be $f_{tT} = 0.33\sqrt{f_{cT}}$, $E_{cT} = 4700\sqrt{f_{cT}}$, and $G_{cT} = E_{cT}/2(1+\upsilon)$, respectively. The refined constitutive model that allows for the tension-stiffening effect presented by Nour at al. [24] is employed by substituting the coefficients at ambient temperature by those calculated at the current temperature. Therefore, the stress–strain relationship for concrete in tension at elevated temperatures is shown in Fig. 6.5 and given by Eqs (6.21a)–(6.21d).

$$\sigma_{tT} = E_{cT}\varepsilon_{tT} \text{ for } 0 \le \varepsilon_{tT} \le \varepsilon_{eT} \tag{6.21a}$$

$$\sigma_{tT} = f_{tT}\left[1 - (1-\beta_T)\frac{\varepsilon_{tT}}{\varepsilon_{1T}}\right] \text{ for } \varepsilon_{eT} < \varepsilon_{tT} \le \varepsilon_{1T} \tag{6.21b}$$

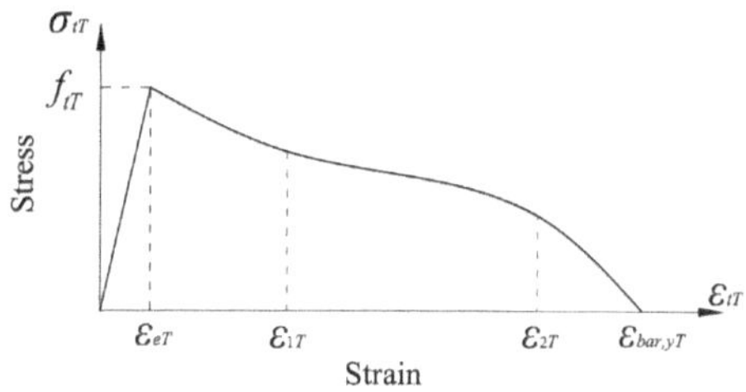

Fig. 6.5 Stress–strain relationship for concrete in tension at elevated temperatures.

$$\sigma_{tT} = \beta_T f_{tT} \text{ for } \varepsilon_{1T} < \varepsilon_{tT} \leq \varepsilon_{2T} \tag{6.21c}$$

$$\sigma_{tT} = \beta_T f_{tT}\left(\frac{\varepsilon_{bar,yT} - \varepsilon_{tT}}{\varepsilon_{bar,yT} - \varepsilon_{2T}}\right) \text{ for } \varepsilon_{2T} < \varepsilon_{tT} \leq \varepsilon_{bar,yT} \tag{6.21d}$$

in which σ_{tT} and ε_{tT} are the concrete tensile stress and strain at temperature T, respectively; ε_{eT} the concrete strain at f_{tT}; $\varepsilon_{bar,\ yT}$ the strain at yielding for steel rebars or the ultimate strain for FRP rebars; β_T the tension-stiffening factor at temperature T given by Eq. (6.22); and ε_{1T} and ε_{2T} are expressed in Eqs (6.23a), (6.23b).

$$\beta_T = exp\left[-1100(\varepsilon_{tT} - \varepsilon_{eT})\left(\frac{\sum E_{bar,T}A_{bar}}{200\sum A_{bar}}\right)\right] \tag{6.22}$$

where $E_{bar,\ T}$ represents the elastic moduli of different reinforcing bars at temperature T and A_{bar} the areas of different reinforcing bars.

$$\varepsilon_{1T} = (1.3 - \beta_T) f_{tT}\left(\frac{1}{E_{bar,T}\rho_{eff}} + \frac{1}{E_{c,T}}\right) \tag{6.23a}$$

$$\varepsilon_{2T} = \varepsilon_{bar,yT} - \beta_T f_{tT}\left(\frac{1}{E_{bar,T}\rho_{eff}} + \frac{1}{E_{c,T}}\right) \tag{6.23b}$$

where ρ_{eff} is the effective ratio of reinforcement.

After cracking, the aggregate interlock and the dowel action of reinforcing bars are accounted for by introducing an appropriate value into the cracked shear modulus G_{cT}'. In the current numerical model, the cracked shear modulus employed by Owen and Figueiras [25], which is assumed to be a function of the current tensile strain, is used as given by

$$G_{cT}' = \begin{cases} 0.25G_{cT}(1 - \varepsilon_{tT}/0.004) & \varepsilon_{tT} < 0.004 \\ 0 & \varepsilon_{tT} \geq 0.004 \end{cases} \tag{6.24}$$

where ε_{tT} is the concrete tensile strain at temperature T.

6.3.2 Temperature-dependent material model of steel

In the current finite element model, the stress–strain relationship recommended by Eurocode 4 [16] for steel in both tension and compression at elevated temperatures is employed. Fig. 6.6 shows the stress–strain relation curve of steel, which can be described by Eqs (6.25a)–(6.25f).

$$\sigma_T = E_{sT}\varepsilon_T \text{ for } \varepsilon_T \le \varepsilon_{pT} \tag{6.25a}$$

$$\sigma_T = \left(f_{pT} - c\right) + \frac{b}{a}\sqrt{a^2 - \left(\varepsilon_{yT} - \varepsilon_T\right)^2} \text{ for } \varepsilon_{pT} < \varepsilon_T \le \varepsilon_{yT}\ (=0.02) \tag{6.25b}$$

$$\sigma_T = \left[\left(f_{uT} - f_{yT}\right)/0.02\right]\varepsilon_T - f_{uT} + 2f_{yT} \text{ for } \varepsilon_{yT}\ (=0.02) < \varepsilon_T < 0.04 \tag{6.25c}$$

$$\sigma_T = f_{uT} \text{ for } 0.04 \le \varepsilon_T \le 0.15 \tag{6.25d}$$

$$\sigma_T = \left[1 - (\varepsilon_T - 0.15)/0.05\right] f_{uT} \text{ for } 0.15 < \varepsilon_T < 0.20 \tag{6.25e}$$

$$\sigma_T = 0 \text{ for } 0.20 \le \varepsilon_T \tag{6.25f}$$

where σ_T and ε_T are the stress and strain of steel at temperature T, respectively, and E_{sT}, f_{pT}, f_{yT}, and f_{uT} the elastic modulus, proportional limit, yield strength, and ultimate strength of steel at temperature T, respectively, determined using the reduction factors k_{ET}, k_{pT}, k_{yT}, and k_{uT}, respectively, as given by Eqs (6.26)–(6.29). Details of these reduction factors are given in Appendix D. ε_{pT} and ε_{yT} correspond to the strains at f_{pT} and f_{yT}, respectively, and the coefficients a, b, and c are given by Eqs (6.30)–(6.32), respectively:

$$E_{sT} = k_{ET}E_s \tag{6.26}$$

$$f_{pT} = k_{pT}f_y \tag{6.27}$$

$$f_{yT} = k_{yT}f_y \tag{6.28}$$

$$f_{uT} = k_{uT}f_y \tag{6.29}$$

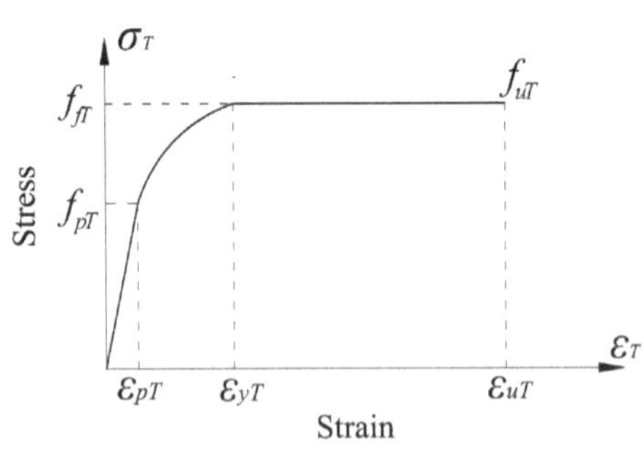

Fig. 6.6 Stress–strain relationship for steel at elevated temperatures [16].

where f_y is the yield strength of steel at ambient temperature

$$a^2 = (\varepsilon_{yT} - \varepsilon_{pT})(\varepsilon_{yT} - \varepsilon_{pT} + c/E_{sT}) \tag{6.30}$$

$$b^2 = E_{sT}(\varepsilon_{yT} - \varepsilon_{pT})c + c^2 \tag{6.31}$$

$$c = \frac{(f_{yT} - f_{pT})^2}{E_{sT}(\varepsilon_{yT} - \varepsilon_{pT}) - 2(f_{yT} - f_{pT})} \tag{6.32}$$

6.3.3 Temperature-dependent material model of FRP

The stress–strain relationship of FRP at elevated temperatures can be assumed to be linear until fracture as shown in Fig. 6.7. The reduction in the mechanical properties of FRP due to high temperatures depends mainly on the specific composition and properties of the material [26]. In this book, Saafi's conservative equations [26] are employed to consider the temperature-dependent material properties of AFRP and GFRP, and the formulas proposed by Rafi et al. [6] are selected for CFRP rebars. The relationships between FRP material properties at ambient and high temperatures are given by Eqs (6.33), (6.34).

$$f_{fT} = k_{fT} f_f \tag{6.33}$$

$$E_{fT} = k_{fET} E_f \tag{6.34}$$

where f_{fT} and E_{fT} are the ultimate tensile strength and modulus of elasticity of FRP rebars at temperature T, respectively; f_f and E_f the ultimate tensile strength and modulus of elasticity of FRP rebars at ambient temperature; and k_{fT} and k_{fET} the temperature reduction factors for tensile strength and modulus of elasticity respectively, which are given by the equations in Appendix E and compared with those of steel rebars in Figs 6.8 and 6.9.

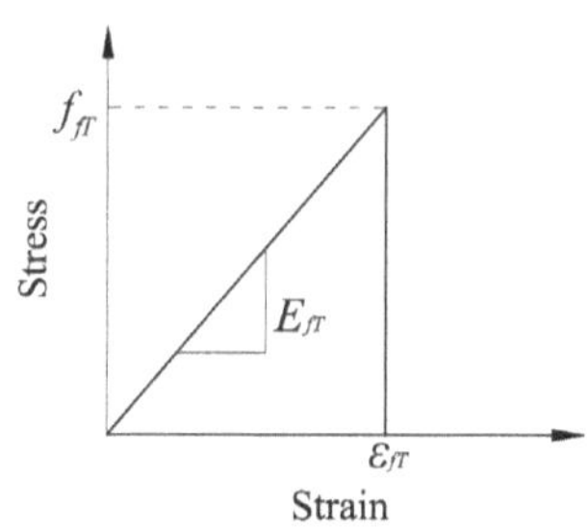

Fig. 6.7 Stress–strain relationship for FRP at elevated temperatures.

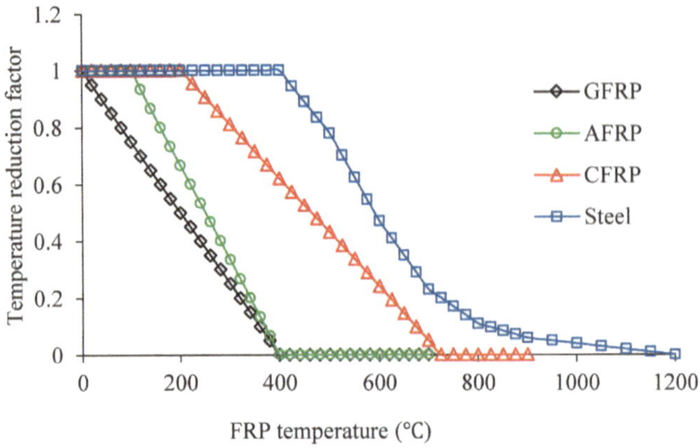

Fig. 6.8 Temperature reduction factors k_{fT} for FRP tensile strength and steel yield strength [6, 16, 26].

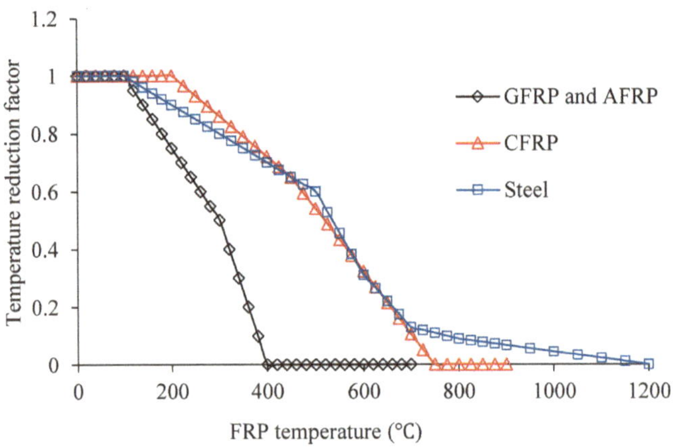

Fig. 6.9 Temperature reduction factors k_{fET} for moduli of elasticity of steel and FRP [6, 16, 26].

6.4 Nonlinear finite element analysis procedures

The nonlinear finite element formulations of the composite beam element have been introduced in Chapter 4. To predict the structural performances of steel-/FRP-reinforced concrete beams at elevated temperatures, the following procedures are taken in the finite element analysis:

(a) Conduct nonlinear structural analysis under mechanical loading at ambient temperature, as described in Chapter 4.

(b) Calculate the thermal load vector $\{F_T^{(e)}\}$ for each quadrilateral element on the cross section of the reinforced concrete beam and then the global thermal load vector.

(c) Divide the temperature exposure time into a number of time steps.

(d) In each time step, carry out the following iterative process.

- Calculate the element conduction matrix $[K_T^{(e)}]$ and mass matrix $[M^{(e)}]$ and then the global conduction matrix and mass matrix.
- Calculate the environmental temperature and the temperature on each node of quadrilateral elements in the current time step. Then, calculate the temperature at the middle point of each quadrilateral element based on the nodal temperatures.
- Check the variations in the material mechanical properties and thermal properties of each quadrilateral element with increases in temperature, and then, update them for the current time step and for each layer.
- Check the variation in thermal strain in each quadrilateral element with increases in temperature, and then, update the thermal strain for the current time step and for each layer.
- Calculate the mechanical strain of each layer, which is the subtrahend of the total strain and thermal strain.
- Conduct nonlinear structural analysis under total mechanical loading at the current temperature. If convergence is achieved, go to the next time step.

6.5 Finite element analysis of reinforced concrete beams at elevated temperatures

In this section, structural behaviours of reinforced concrete beams, including a steel-reinforced concrete beam and a FRP-reinforced concrete beam, under fire conditions are predicted as numerical examples. The 1-D composite beam element and the nonlinear finite element analysis procedures introduced in this chapter are employed in the numerical simulations.

6.5.1 A steel-reinforced concrete beam under fire condition

A steel-reinforced concrete beam tested by Dotreppe and Franssen [27] is analysed in this example. Eight composite beam elements are used in the current finite element model to discretise this reinforced concrete beam. The beam, of length 6500 mm, width 200 mm, and depth 600 mm, is simply supported and reinforced with three 22-mm-diameter steel bars in the bottom area and two 12-mm-diameter steel bars in the compression zone. The compressive strength of concrete and the yield strength of reinforcing bars are 15 and 300 MPa, respectively. A four-point bending load of 65 kN is applied on the top of the beam, and the thermal loading is applied according to the ISO R834 standard fire curve. Details of the cross section of the steel-reinforced concrete beam and its loading are shown in Fig. 6.10. The time-dependent temperature curve for the central reinforcing bar obtained from the heat transfer model and the growth in the midspan deflection with increases in temperature obtained from the finite element model are compared with the tested results [27] in Figs 6.11 and 6.12, respectively.

It can be seen from Fig. 6.11 that the predicted temperature curve for the central reinforcing bar is fairly accurate in the entire time–temperature history when comparing with the test data [27]. The temperature–deflection relationship shown in Fig. 6.12 also illustrates a good agreement between the present numerical prediction and the experimental data [27]. The predicted failure of the steel-reinforced concrete beam

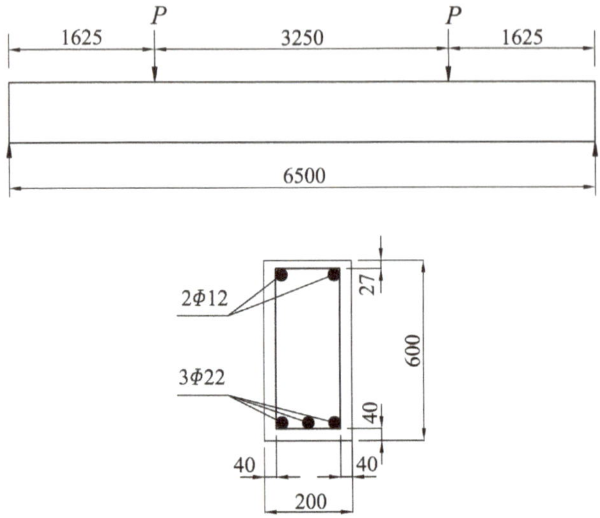

Fig. 6.10 Cross section and loading system of the steel-reinforced concrete beam (unit: millimeter) [27].

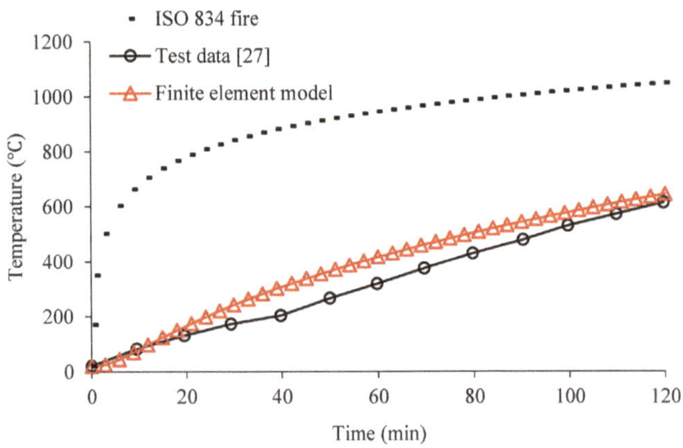

Fig. 6.11 Temperature curves for the central reinforcing bar of the steel-reinforced concrete beam.

occurs at a temperature of around 1020°C, which also agrees with the results from experimental test [27].

6.5.2 A FRP-reinforced concrete beam under fire condition

A CFRP-reinforced concrete beam tested by Rafi et al. [3] under the ISO R834 standard heating curve is modelled in this section using the composite beam element. The cross section of the beam is 120 × 200 mm, and the total length and span length

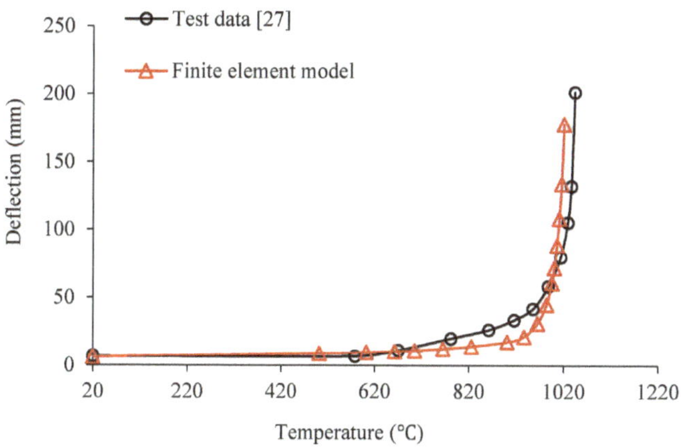

Fig. 6.12 Temperature–deflection relationships at midspan of the steel-reinforced concrete beam under fire condition.

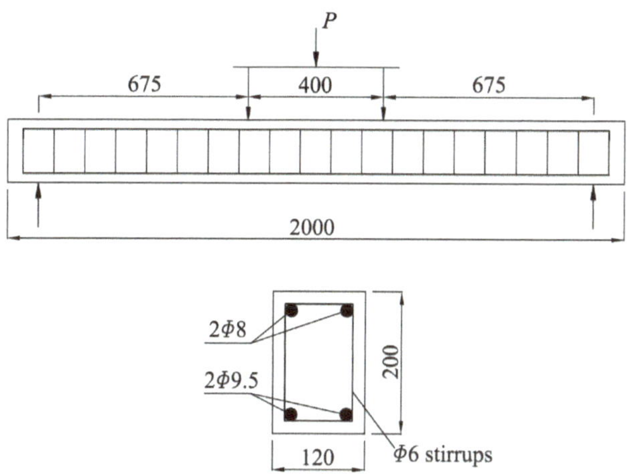

Fig. 6.13 Details of CFRP-reinforced concrete beam (unit: millimeter) [3].

of the beam are 2000 and 1750 mm, respectively. The beam is reinforced with two longitudinal 9.5-mm-diameter CFRP bars in the tension zone and two 8-mm-diameter steel bars in the compression area, and a 20-mm concrete cover is used all around the beam. The ultimate strength and elastic modulus of CFRP reinforcing bars are 1676 MPa and 135.9 GPa, respectively, whilst, for steel reinforcing bars in compression, the yield strength and elastic modulus are 566 MPa and 194 GPa, respectively. The concrete compressive strength is 33.22 MPa. Forty percent of the ultimate load at room temperature (about 30 kN) is applied on the top of the beam. Details of the cross section and loading setup for the CFRP-reinforced concrete beam are shown in Fig. 6.13.

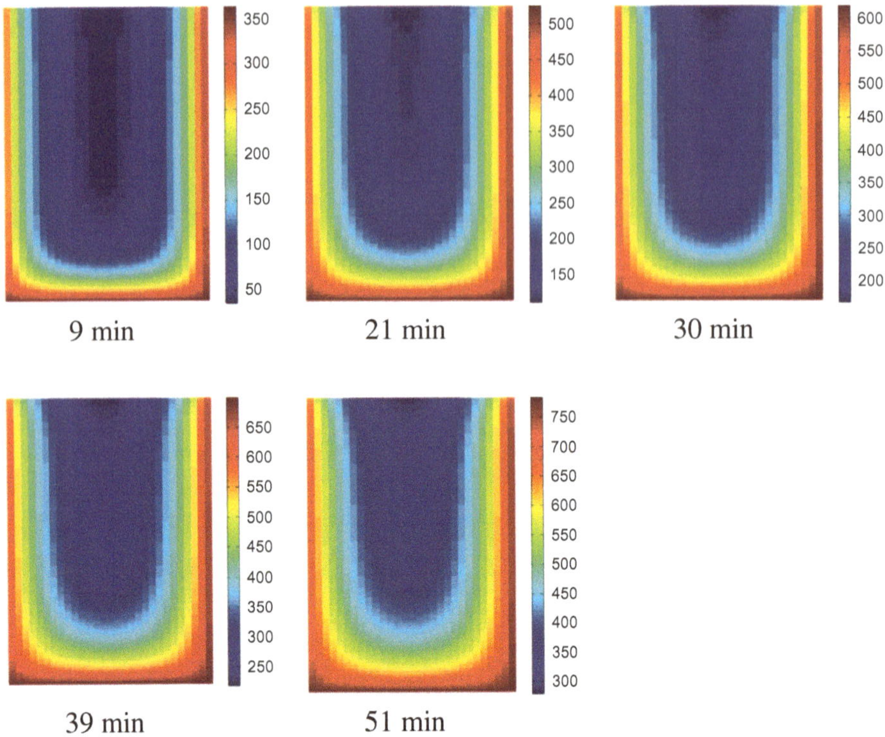

Fig. 6.14 Temperature distributions across the cross section of CFRP-reinforced concrete beam at different fire exposure times.

The temperature distributions across the cross section at different fire exposure times are predicted and shown in Fig. 6.14. The tested and calculated temperature versus time relationships for the surface of one of the CFRP rebars are compared in Fig. 6.15. As can be seen, the overall trend of the time–temperature curve obtained from the present finite element model is the same as the tested curve, and the present numerical prediction is also in a very good agreement with that obtained from Rafi et al.'s 3-D model [6]. Fig. 6.16 shows the comparison of the temperature–deflection relationships of the CFRP-reinforced concrete beam obtained from the present model, the experimental test [3], and the 3-D finite element model proposed by Rafi et al. [6]. It can be seen that both Rafi et al.'s 3-D model [6] and the present finite element model established using the 1-D composite beam element match very well with the test data [3], and compared with the 3-D model, the current numerical model is much more computationally efficient and with even higher accuracy. The failure of the CFRP-reinforced concrete beam is predicted to occur after 54 min of heating, which is also close to the 51 min observed in the test [3].

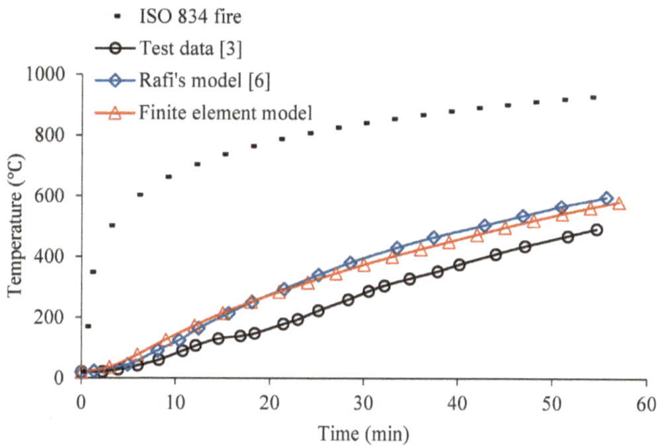

Fig. 6.15 Time–temperature curves on the surface of CFRP rebar.

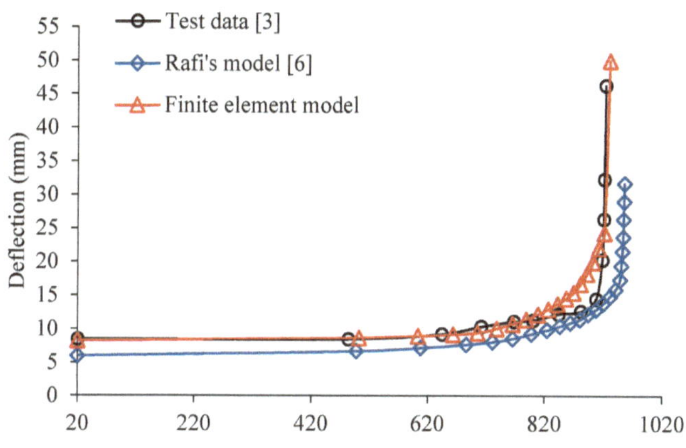

Fig. 6.16 Temperature–deflection relationships at midspan of the CFRP-reinforced concrete beam under fire condition.

References

[1] V.K.R. Kodur, L.A. Bisby, S.H.C. Foo, Thermal behavior of fire-exposed concrete slabs reinforced with fiber-reinforced polymer bars, ACI Struct. J. 102 (6) (2005) 799–807.

[2] M. Elbadry, O. Elzaroug, Control of cracking due to temperature in structural concrete reinforced with CFRP bars, Compos. Struct. 64 (1) (2004) 37–45.

[3] M.M. Rafi, A. Nadjai, F. Ali, Fire resistance of carbon FRP reinforced-concrete beams, Mag. Concr. Res. 59 (4) (2007) 245–255.

[4] M.M. Rafi, A. Nadjai, Behavior of hybrid (steel-CFRP) and CFRP bar-reinforced concrete beams in fire, J. Compos. Mater. (2010) 1–12.

[5] M.M. Rafi, A. Nadjai, Experimental behaviour of carbon FRP reinforced concrete beams at ambient and elevated temperatures, J. Adv. Concr. Technol. 6 (3) (2008) 431–441.

[6] M.M. Rafi, A. Nadjai, F. Ali, Finite element modeling of carbon fiber-reinforced polymer reinforced concrete beams under elevated temperatures, ACI Struct. J. 105 (6) (2008) 701–710.
[7] X. Lin, Y.X. Zhang, Nonlinear finite element analyses of steel/FRP-reinforced concrete beams in fire conditions, Compos. Struct. 97 (2013) 277–285.
[8] E.G. Nawy, G.E. Neuwerth, C.J. Phillips, Behavior of fiber glass-reinforced concrete beams, J. Struct. Div. 97 (9) (1971) 2203–2215.
[9] E.G. Nawy, G.E. Neuwerth, Fiberglass reinforced concrete slabs and beams, J. Struct. Div. 103 (2) (1977) 421–440.
[10] H. Saadatmanesh, M.R. Ehsani, Fiber composite bar for reinforced concrete construction, J. Compos. Mater. 25 (2) (1991) 188–203.
[11] B. Benmokrane, O. Chaallal, R. Masmoudi, Glass fibre reinforced plastic (GFRP) rebars for concrete structures, Constr. Build. Mater. 9 (6) (1995) 353–364.
[12] M.M. Rafi, A. Nadjai, Behavior of hybrid (steel-CFRP) and CFRP bar-reinforced concrete beams in fire, J. Compos. Mater. 45 (15) (2011) 1573–1584.
[13] M.M. Rafi, A. Nadjai, Fire tests of hybrid and carbon fiber-reinforced polymer bar reinforced concrete beams, ACI Mater. J. 108 (3) (2011) 252–260.
[14] European Committee for Standardization, in: M. Keynes (Ed.), Eurocode 2: Design of concrete structures. Part 1–2. General rules-Structural Fire Design (DD ENV 1992-1-2), British Standards Institution, 1995.
[15] F. Ceroni, E. Cosenza, M. Gaetano, M. Pecce, Durability issues of FRP rebars in reinforced concrete members, Cem. Concr. Compos. 28 (10) (2006) 857–868.
[16] European Committee for Standardization, Eurocode 4: Design of Composite Steel and Concrete Structures. Part 1–2. General rules-Structural fire design (BS EN 1994-1-2), 2005.
[17] M.M. Rafi, A. Nadjai, F. Ali, P. O'Hare, Evaluation of thermal resistance of FRP reinforced concrete beams in fire, J. Struct. Fire Eng. 2 (2) (2011) 91–107.
[18] J. Kruppa, B. Zhao, Fire resistance of composite beams to Eurocode 4 Part 1.2, J. Constr. Steel Res. 33 (1-2) (1995) 51–69.
[19] R.W. Lewis, K. Morgan, H.R. Thomas, K.N. Seetharamu, The Finite Element Method in Heat Transfer Analysis, John Wiley & Sons Ltd., Chichester, 1996.
[20] Z. Nizamuddin, B. Bresler, Fire response of reinforced concrete slabs, J. Struct. Div. 105 (8) (1979) 1653–1671.
[21] B. Ellingwood, T.D. Lin, Flexure and shear behavior of concrete beams during fires, J. Struct. Eng. 117 (2) (1991) 440–458.
[22] K.D. Hertz, Concrete strength for fire safety design, Mag. Concr. Res. 57 (8) (2005) 445–453.
[23] M.A. Youssef, M. Moftah, General stress–strain relationship for concrete at elevated temperatures, Eng. Struct. 29 (10) (2007) 2618–2634.
[24] A. Nour, B. Massicotte, E. Yildiz, V. Koval, Finite element modeling of concrete structures reinforced with internal and external fibre-reinforced polymers, Can. J. Civ. Eng. 34 (3) (2007) 340–354.
[25] D.R.J. Owen, J.A. Figueiras, E. Hinton, D.R.J. Owen (Eds.), Ultimate load analysis of reinforced concrete plates and shells including geometric nonlinear effects, in: Finite Element Software for Plates and Shells, Pineridge Press, Swansea, UK, 1984.
[26] M. Saafi, Effect of fire on FRP reinforced concrete members, Compos. Struct. 58 (1) (2002) 11–20.
[27] J.C. Dotreppe, J.M. Franssen, The use of numerical models for the fire analysis of reinforced concrete and composite structures, Eng. Anal. 2 (2) (1985) 67–74.

Finite element analysis of FRP-strengthened reinforced concrete beams under static and cyclic loads

7

7.1 Introduction

FRP strengthening system has been demonstrated to be feasible and efficient in improving flexural stiffness and strength of reinforced concrete (RC) beams [1]. The life of FRP-strengthened RC beams can be increased by three times of that of unstrengthened RC beams under cyclic load, and the fatigue strength can be increased by 15%–30% [2]. RC beams strengthened by FRP have also been found to exhibit extended fatigue endurance limit when subjected to repetitive cyclic loading [3].

A large number of experimental studies have been conducted to investigate the flexural behaviour of FRP-strengthened RC beams. Although experimental tests provide the source of reliable data and results, it is usually very time and resource consuming, especially for tests under cyclic loading. At present, numerous numerical studies on the FRP-strengthened RC beams have been reported; however, the majority have been focused on the static behaviour with a perfect bond assumption between FRP, adhesive, and concrete. In practice, the structural performance of FRP-strengthened RC beams depends not only on its components such as concrete, steel, and FRP but also on the bond behaviour between FRP and concrete. The bond between concrete and FRP laminate plays a significant role in transferring stress from the former to the latter, and the bond behaviour is a critical aspect for the structural performance of FRP-strengthened RC beams [4]. For an accurate numerical modelling of the structural behaviour of FRP-strengthened RC beams, appropriate material models for each component and the bond behaviour are essential.

Consideration of bond–slip effect between FRP, adhesive, and concrete is utmost important to obtain an accurate performance-based prediction of structural behaviour under cyclic loads. With the application of cyclic loads, the material properties of structural components are degraded with the number of cycles, and the interfacial behaviour between the components is also degraded, which would affect the structural behaviour of FRP-strengthened RC beams [5]. Owing to the complex mechanisms induced by the heterogeneity of materials and the various parameters that influence the flexural fatigue life, it is still difficult to accurately predict the flexural fatigue response of RC beams strengthened with FRP [2]. At present, a number of finite element analyses have been reported. For example, Loo et al. [6] simulated CFRP-strengthened beam under cyclic loading. The stress–strain relationship for concrete in compression under fatigue loading was simulated based on the model proposed by Holmen [7], the steel reinforcement was modelled based on the CEB-FIP model code 1993 [8], and the elastic–brittle stress–strain relationship was used for

Nonlinear Finite Element Analysis of Composite and Reinforced Concrete Beams
https://doi.org/10.1016/B978-0-12-816899-8.00007-9

FRP, assuming that the strength of FRP would not change with fatigue. The 1-D bond interface element developed by Chong et al [9] was used to describe bond–slip between concrete and FRP interface. However, the degradation of FRP properties with the number of cycles was neglected in this study. Ferrier et al. [10] modelled CFRP-strengthened RC beam under cyclic loading. The bond–slip between concrete, adhesive, and FRP was modelled based on the analytical solution provided by Täljsten [11]. The degradation of concrete properties with the number of loading cycles was simulated based on material model proposed by Aas-Jakobsen [12], and the decrease in Young's modulus was simulated based on Balaguru and Shah's model [13]. The reinforcing steel was simulated using the model proposed by Mander et al. [14]. The response of FRP and adhesive joint subjected to cyclic load was described based on the Ferrier et al.'s model [10].

In this chapter, four finite element models denoted as FEM-P and FEM-B for static loading condition [15] and FEM-PC and FEM-BC for cyclic loading condition [16, 17] are introduced for accurate and efficient prediction of structural behaviour of FRP-strengthened RC beams. FEM-P and FEM-PC are the models with perfect bond assumption, and FEM-B and FEM-BC are the models with bond–slip effect. For FRP-strengthened RC beams under cyclic loads, the deterioration of material properties with the number of loading cycles is included. Furthermore, FEM-B and FEM-BC include the responses of interfaces between FRP/adhesive/concrete with the number of loading cycles. Adhesive is an important component for FRP-strengthened RC beams. Pathak and Zhang [16] and Molina et al. [18] claimed that the effect of adhesive could not be neglected whilst developing a FE model, because the adhesive could be susceptible to damage or plasticity. In this chapter, the effect of bond–slip on the structural performance is demonstrated by comparing the numerical results obtained from the FE models with and without bond–slip.

7.2 Finite elements in numerical models

Four different types of elements are used in the development of finite element models, including solid element for concrete and adhesive, link element for steel reinforcement, shell element for FRP, and spring element for interface interaction. They are described in the following sections.

7.2.1 Concrete

In the FE model, SOLID65 element is used to model concrete, which is an eight-node three-dimensional solid element with three degrees of freedom at each node, that is, translations in x, y, and z directions. This element can simulate cracking under tension in three orthogonal directions, crushing under compression, plastic deformation, and creep. In this element, the cracks in concrete are regarded as smeared bands at the element integration points when the principal stress exceeds the concrete tensile strength and the material properties are adjusted accordingly. The geometry of SOLID65 element is shown in Fig. 7.1.

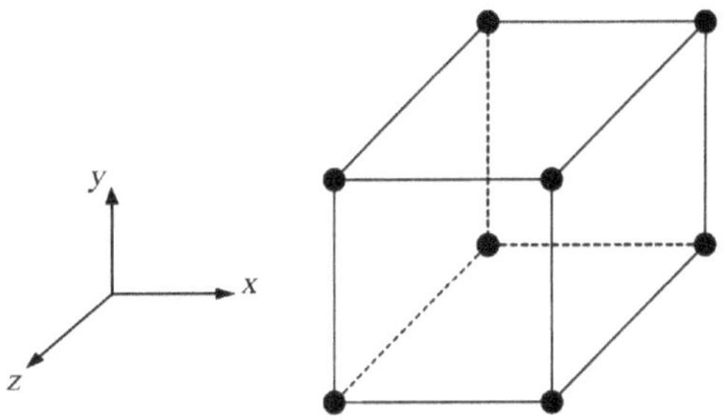

Fig. 7.1 Geometry of SOLID65 element.

7.2.2 Steel

LINK180 is a 3-D spar with three degrees of freedom at each of the two nodes, that is, translations in x, y, and z directions. This element can be used to model trusses, sagging cables, links, springs, and so on. It is used to model steel reinforcements in this analysis, as embedded in solid mesh. This option is favoured over the alternative smeared stiffness capability in solid element as it allows the reinforcement to be precisely located whilst maintaining a relatively coarse mesh for the surrounding concrete medium [19]. The geometry for LINK180 element is shown in Fig. 7.2.

7.2.3 FRP

SHELL181 element is used to model FRP sheet. In a shell element, the thickness is relatively small compared with the other dimensions of the element. Also, no additional nodes need to be generated, and the aspect ratio of the element does not exceed the acceptable limit; hence, shell element is always preferred for the modelling of FRP [20]. SHELL181 element is a four-node element with six degrees of freedom at each node: translations in x, y, and z directions and rotations about x-, y-, and z-axes. In this analysis, the membrane only option is used, so this element doesn't have bending stiffness, and hence, each node of the element has only three degrees of freedom, that is, translations in x, y, and z directions. The element geometry for the SHELL181 element is shown in Fig. 7.3.

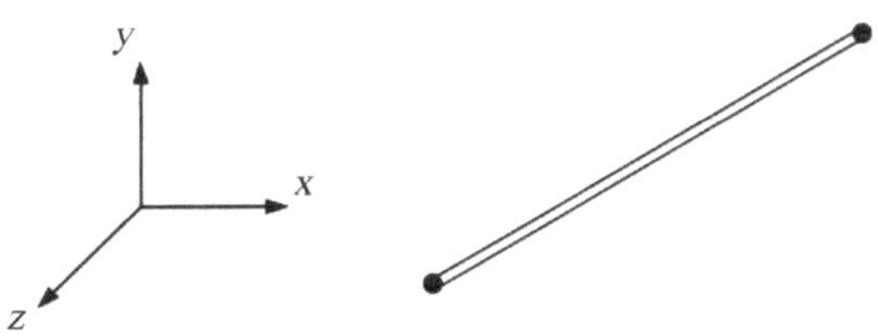

Fig. 7.2 Geometry of LINK180 element.

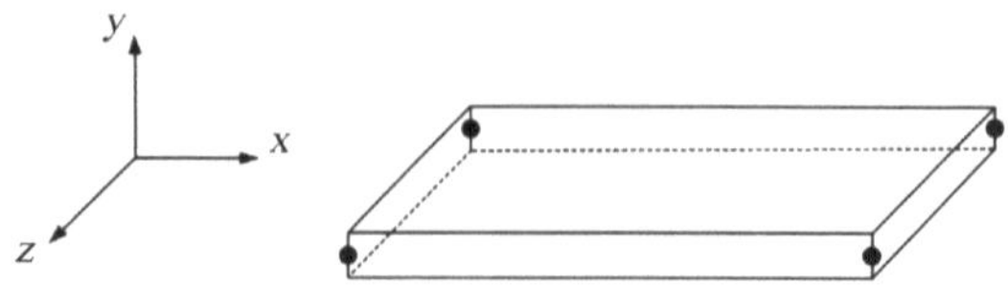

Fig. 7.3 Geometry of SHELL181 element.

7.2.4 Adhesives

The adhesive is modelled using SOLID45 element. It is an eight-node three-dimensional element having three degrees of freedom at each node, that is, translations in x, y, and z directions. The element has plasticity, creep, swelling, stress stiffening, large deflection, and large strain capabilities. The element consists of two sets of nodes. In this analysis, one set of nodes is used with concrete element, and the other set is used with FRP element. The geometry of the SOLID45 element is shown in Fig. 7.4.

7.2.5 Concrete/adhesive/FRP interfaces

Amongst different debonding failure modes, such as intermediate crack debonding, concrete cover separation, end peeling, cohesive failure in concrete, adhesion failure in concrete and adhesive, and adhesion failure in adhesive and FRP, the latter three are dominant [21]. In this analysis, the behaviour of bond between FRP, adhesive, and concrete is modelled by introducing spring elements at the interfaces. The nodes of concrete and adhesive that align at the same locations are connected by these interface elements. This method is also applied for the interface between adhesive and FRP.

The unidirectional spring element COMBIN39 with a nonlinear generalised force–deflection capability is used as the interface element. The element has longitudinal or torsional capability in one-, two-, or three-dimensional applications. The longitudinal option is a uniaxial tension–compression element with up to three degrees of freedom at each node, that is, translations in x, y, and z directions. In this analysis, bending and torsion are not considered. The geometry of the COMBIN39 element is shown in Fig. 7.5. The cross section of a FRP-strengthened RC beam is shown in Fig. 7.6, and the geometry of the interface in the finite element model is illustrated in Fig. 7.7.

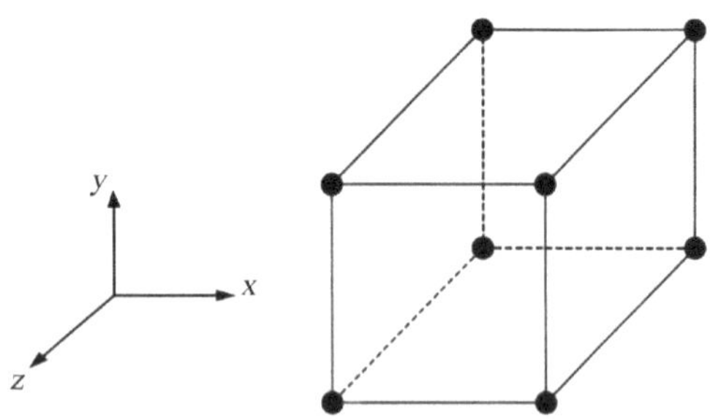

Fig. 7.4 Geometry of SOLID45 element.

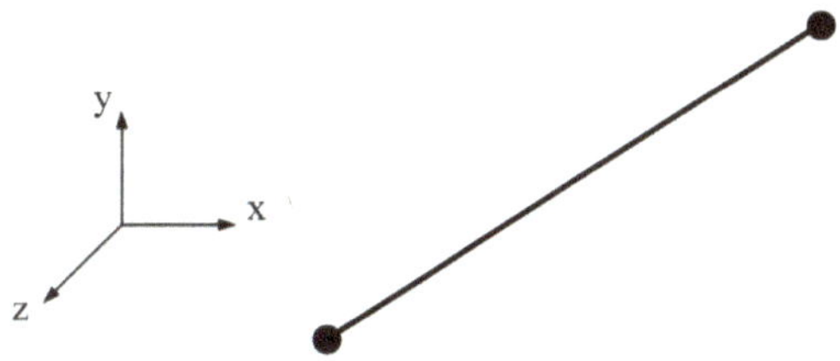

Fig. 7.5 Geometry of COMBIN39 element.

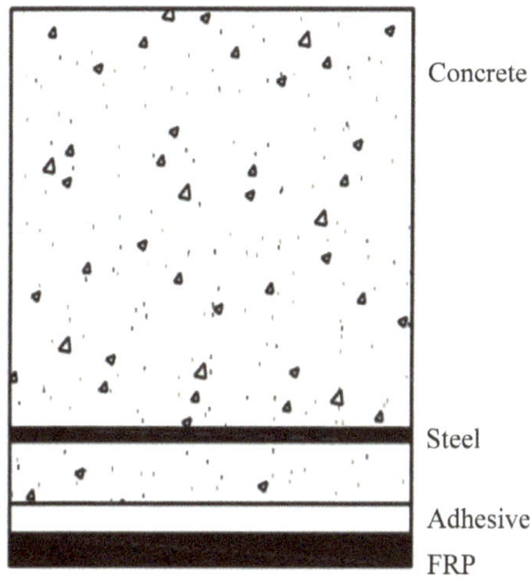

Fig. 7.6 Cross section of FRP-strengthened RC beam.

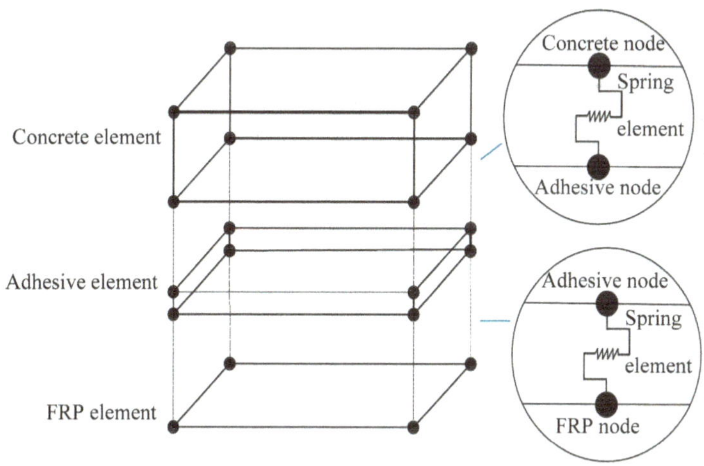

Fig. 7.7 Interface geometry in FEM-BC.
From P. Pathak, Y.X. Zhang, Numerical study of structural behavior of fiber-reinforced polymer-strengthened reinforced concrete beams with bond-slip effect under cyclic loading, Struct. Concr. 3 (2018) 1–11. With permission from John Wiley and Sons.

For FEM-P and FEM-PC, since perfect bond is assumed between the interfaces, spring element in Fig. 7.7 is not included.

7.3 Material models

For accurate prediction of FRP-strengthened RC beams, it is very essential to employ appropriate material models for different components, that is, concrete, steel, FRP, and adhesive. The material models employed in the present numerical model are introduced in the following sections.

7.3.1 Concrete

The compressive behaviour of concrete is modelled based on the nonlinear stress–strain relationship proposed by Popovics [22] and later modified by Thorenfeldt et al. [23]. The stress–strain curve is shown in Fig. 7.8A and described by Eqs (7.1)–(7.4). This model is also applicable for high-strength concrete where the postpeak branch becomes steeper with the increase of the concrete compressive strength [23]:

$$\sigma_c = \frac{n f_c' (\varepsilon_c/\varepsilon_0)}{n - 1 + (\varepsilon_c/\varepsilon_0)^{nk}} \tag{7.1}$$

$$n = 0.8 + f_c'/17 \tag{7.2}$$

$$k = \left(0.67 + f_c'/62\right) > 1 \text{ for } 1 < (\varepsilon_c/\varepsilon_0) \leq \varepsilon_{cu} \tag{7.3}$$

$$k = 1 \text{ for } (\varepsilon_c/\varepsilon_0) \leq 1 \tag{7.4}$$

where f_c' is the compressive strength of concrete and σ_c and ε_c the concrete compressive stress and compressive strain, respectively. ε_{cu} is the ultimate compressive strain of concrete, and ε_0 the corresponding compressive strain at the compressive strength, which is calculated by Eq. (7.5) [25].

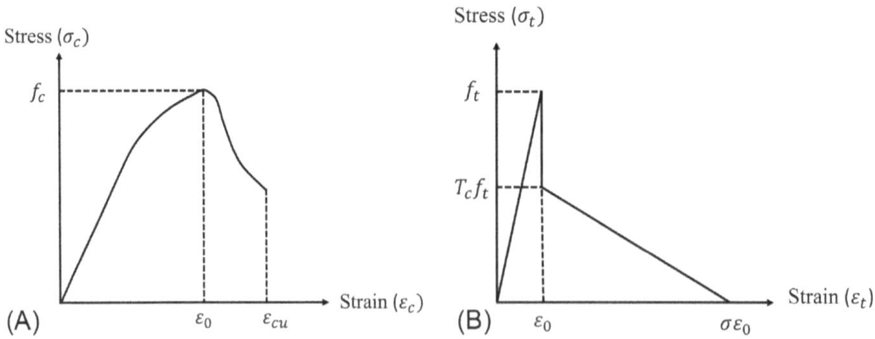

Fig. 7.8 Stress–strain relationship of concrete: (A) under compression [23] and (B) under tension [24].

$$\varepsilon_0 = 1.71 \times f_c'/E_c \quad (7.5)$$

in which E_c is the Young's modulus of concrete.

As per ACI 318 code [26], the Young's modulus (E_c) and the tensile strength (f_t) of concrete are given by

$$E_c = 4700\sqrt{f_c'} \quad (7.6)$$

$$f_t = 0.62\sqrt{f_c'} \quad (7.7)$$

For concrete in tension, the stress–strain curve is assumed to be isotropic and linearly elastic up to the maximum tensile strength (f_t), after which concrete cracks and the strength gradually reduces to zero as shown in Fig. 7.8B, in which T_c is the multiplier for the amount of tensile stress relaxation with a default value of 0.6 in ANSYS.

The failure of concrete under multiaxial stress state is considered based on the William-Warnke yield criterion [27]. Cracking is assumed to occur when the principal tensile stress in any direction lies outside the failure surface, after which the elastic modulus of concrete is set to zero in the direction parallel to it.

The input data required in ANSYS for describing the material properties of concrete include Poisson's ratio (υ), elastic modulus (E_c), uniaxial tensile stress (f_t), uniaxial compressive stress (f_c'), and shear transfer coefficient (β_t). The value of β_t varies from zero (smooth crack) to one (rough crack). β_t is used to determine how much shear force can be transferred across an open or closed crack. In this model, β_t is assumed to be 1 for a closed crack and 0.3 for an open crack.

7.3.2 Steel

Steel is assumed to be isotropic and elastic-perfectly plastic, which behaves identically in tension and compression with stiffness only in the axial direction. A typical stress–strain relationship of steel is shown in Fig. 7.9.

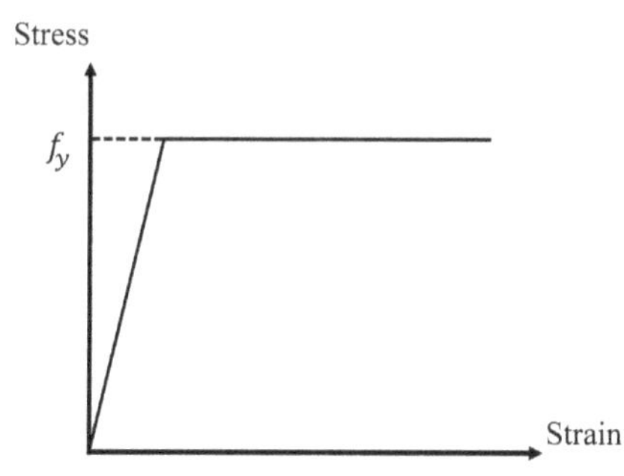

Fig. 7.9 Stress–strain relationship of steel.

7.3.3 FRP

In the study reported by Shrestha [20], as properties of FRP in the transverse direction were about 1% of the properties along the fibre orientation, the FRP transverse properties were assumed to be zero in the stiffness matrix. Similarly, FRP was assumed as isotropic and orthotropic material by Obaidat [28]. In the current FE model, FRP is assumed to be isotropic and linearly elastic with material properties along the fibre direction only. Once the tensile strength of FRP is reached, FRP rupture occurs.

7.3.4 Adhesive

Generally speaking, an adhesive is required to bind two materials to avoid their separation. The use of adhesives offers many advantages over other binding techniques such as sewing, mechanical fastening, and thermal bonding, which include the ability to bind different materials together and distribute stress more efficiently across the joint, cost effectiveness for an easily mechanised process, improvement in aesthetic design, and increased design flexibility [29]. In the present FE model, the adhesive is assumed to be linear and elastic. A typical stress–strain relationship is shown in Fig. 7.10.

7.4 Bond–slip model

Amongst various bond stress–slip models, the bilinear model proposed by Lu et al. [30] is widely accepted (Fig. 7.11), which can give accurate prediction of bond–slip behaviour [4, 31]. This model is adopted herein to simulate the bond–slip behaviour at the interfaces between FRP, adhesive, and concrete and given by the following equations:

$$\tau = \tau_{max} \frac{s}{s_0} \text{ if } s \leq s_0 \tag{7.8}$$

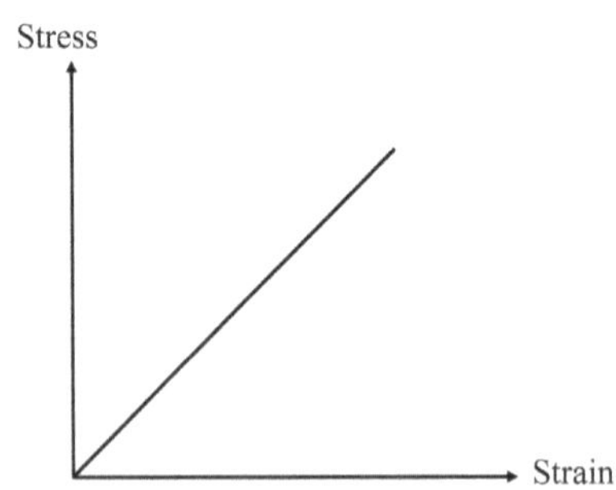

Fig. 7.10 Stress–strain relationship of adhesive.

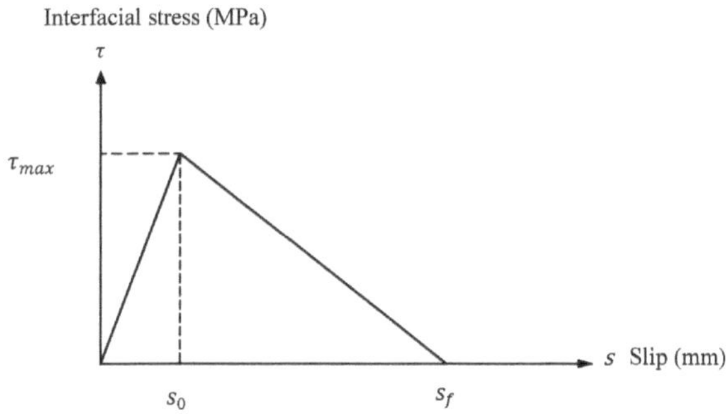

Fig. 7.11 Bilinear model proposed by Lu et al. [30].

$$\tau = \tau_{max} \frac{s_f - s}{s_f - s_0} \text{ if } s_0 < s \leq s_f \tag{7.9}$$

$$\tau = 0 \text{ if } s > s_f \tag{7.10}$$

$$s_f = 2G_f / \tau_{max} \tag{7.11}$$

$$s_0 = 0.0195 \beta_w f_t \tag{7.12}$$

$$\tau_{max} = 1.5 \beta_w f_t \tag{7.13}$$

$$G_f = 0.308 \beta_w^2 \sqrt{f_t} \tag{7.14}$$

$$\beta_w = \sqrt{\frac{2.25 - \frac{b_f}{b_c}}{1.25 + \frac{b_f}{b_c}}} \tag{7.15}$$

where τ is the local bond stress, τ_{max} the maximum local bond stress, s the local bond–slip, s_0 the local bond–slip at τ_{max}, s_f the local slip when bond stress τ reduces to zero, G_f the interfacial fracture energy, β_w the width ratio factor, f_t the concrete tensile strength, b_f the width of FRP plate, and b_c the width of concrete.

7.5 Material behaviours of concrete, steel, and FRP under cyclic load

7.5.1 Concrete

In the current numerical model, a material model proposed by Zanuy et al. [32] is employed for the modelling of concrete under cyclic load. The maximum strain of concrete under cyclic load and the stress redistribution in concrete are included. Also, this model is capable of including a continuous change in stress and strain, and the failure of concrete from 80% to 100% fatigue life is covered. This material model is given by Eqs (7.16)–(7.19a)–(7.19f). The decrease in Young's modulus under cyclic load is based on the approximation of Holmen's model [7] using piecewise linear approach, which has been successfully used by Loo [33]:

For $0.0 \le \frac{N}{N_f} \le 0.1$:

$$\frac{\varepsilon_{max}}{\varepsilon_{ini}} = 1 + A\frac{N}{N_f} + B\left(\frac{N}{N_f}\right)^2 \tag{7.16}$$

For $0.1 \le \frac{N}{N_f} \le 0.8$:

$$\frac{\varepsilon_{max}}{\varepsilon_{ini}} = \varepsilon_{1-2} + \varepsilon_2\left(\frac{N}{N_f} - 0.1\right) \tag{7.17}$$

For $0.8 \le \frac{N}{N_f} < 1$:

$$\frac{\varepsilon_{max}}{\varepsilon_{ini}} = \varepsilon_{1-2} + \varepsilon_2\left(\frac{N}{N_f} - 0.1\right) + C\left(\frac{N}{N_f} - 0.8\right)^2 \tag{7.18}$$

where N is the number of cycles and N_f the number of cycles at failure. ε_{ini} is the initial strain at the first load cycle ($N = 1$), and

$$A = 20(\varepsilon_{1-2} - 1) - \varepsilon_2 \tag{7.19a}$$

$$B = 100(1 - \varepsilon_{1-2}) + 10\varepsilon_2 \tag{7.19b}$$

$$C = 25\left(\frac{\varepsilon_{fail}}{\varepsilon_{ini}} - \varepsilon_{1-2} - 0.9\varepsilon_2\right) \tag{7.19c}$$

$$\varepsilon_{1-2} = \frac{1.184}{S_{max}} \tag{7.19d}$$

$$\varepsilon_2 = \frac{0.74037}{S_{max}} \tag{7.19e}$$

$$\varepsilon_2 \le \frac{1}{0.9}\left(\frac{\varepsilon_{fail}}{\varepsilon_{ini}} - \varepsilon_{1-2}\right) \tag{7.19f}$$

where S_{max} is the ratio of maximum stress to concrete strength and ε_{fail} the strain at N_f.

Once the maximum strain is obtained, the Young's modulus of concrete after N loading cycles can be estimated using Eq. (7.20):

$$E_N = E_c\left(1 - 0.44\frac{N}{N_f}\right) \tag{7.20}$$

where E_c is the initial Young's modulus of concrete.

In the present model, a subroutine entitled userconcrete.f (see Appendix J) is used to define this material model for considering the variations in the concrete material properties under cyclic load.

7.5.2 Steel

For the material properties of steel under cyclic load, the modified universal slope method developed by Muralidharan and Manson [34] is employed. The strain of steel is expressed as a function of the number of cycles, the steel ultimate strength, and the ultimate strain as

$$\varepsilon_{sn} = 0.623\left(\frac{f_u}{E_s}\right)^{0.832}\left(2N_f\right)^{-0.09} + 0.0196(\varepsilon_{su})^{0.155}\left(\frac{f_u}{E_s}\right)^{-0.53}\left(2N_f\right)^{-0.56} \tag{7.21}$$

where ε_{sn} is the strain of steel, f_u the ultimate stress, E_s the Young's modulus of steel, N_f the number of loading cycles, and ε_{su} the ultimate strain of steel.

The presence of the f_u/E_s term in the coefficient of the plastic line implies a role of the ultimate tensile strength and elastic modulus in the very-low-cycle fatigue range. This model is very helpful to the analysis and design for low alternating strain amplitudes and very high cyclic life [34].

A subroutine called usersteel.f (see Appendix K), which is developed based on the user subroutine function Usermat.f in ANSYS, is employed to describe the material behaviour of steel.

7.5.3 FRP

For FRP, the equation proposed by Ferrier et al. [10] is used, which explains how the Young's modulus of FRP degrades with the number of loading cycles. The relationship between FRP Young's modulus and the number of loading cycles N is expressed as

$$E_{fn} = m - n \times \log(N) \tag{7.22}$$

where m (in N/mm^3) is the Young's modulus of FRP under static loading and $n = 1100$.

In the FE model, a subroutine called userfrp.f (see Appendix L) is employed to define the Young's modulus of FRP under cyclic loads.

7.5.4 Bond stress–slip relationship

In the current FE model, the bond stress–slip relationship proposed by Loo et al. [6], as given by Eqs (7.23), (7.24), is adopted to represent the behaviour of FRP/adhesive/concrete bonded interfaces subjected to cyclic load.

$$S_{max} = \frac{\tau_{ave,}\max}{E_{b0}}\left[1 + a(logN)^b\left(\frac{\Delta\tau_{ave}}{\tau_{ave,f}}\right)^c\right] \tag{7.23}$$

$$E_b = E_{b0}\left[1 + \alpha(logN)^\beta\left(\frac{\Delta\tau_{ave}}{\tau_{ave,f}}\right)^\gamma\right] \tag{7.24}$$

where S_{max} (mm) is the maximum slip, E_{b0} (N/mm^3) the bond modulus at the first cycle, E_b (N/mm^3) the bond modulus at the Nth cycle, $\Delta\tau_{ave}$ the bond stress range, and $\tau_{ave,\ f}$ the average bond stress at failure. The unit of $\tau_{ave,\ max}$ (maximum applied bond stress) is megapascal. The values of a, b, c, α, β, and γ are 1.068, 1.402, 20612, −190.3, 0.99, and 8.797, respectively.

7.6 Loading conditions

The cyclic load is defined as a sinusoidal load with the maximum and minimum amplitudes of F_{max} and F_{min} and changes with a fixed frequency. The sinusoidal load is represented by

$$F = ksin(at) + l \tag{7.25}$$

where $\kappa = \frac{F_{max} - \mathrm{F_{min}}}{2}$, $l = \frac{F_{max} + F_{min}}{2}$, and t is the loading time.

7.7 Finite element analysis of FRP-strengthened RC beams

7.7.1 CFRP-strengthened RC beam under static load

In this section, a CFRP-strengthened RC beam tested by Arduini et al. [35] is analysed using FEM-P and FEM-B. The RC beam is under four-point bending load. The geometric configurations of the beam, the steel reinforcements, and the CFRP strengthening plate are shown in Fig. 7.12. The bottom face of the RC beam is externally strengthened with 1000-mm-long, 300-mm-wide, and 1.2-mm-thick CFRP plate using epoxy adhesive. The material properties of concrete, steel, FRP, and epoxy are given in Table 7.1. Due to symmetry, only a quarter of the beam is analysed. The numerical results obtained from FEM-B with bond–slip effect are also compared with those obtained from FEM-P with perfect bond assumption.

A convergence test is firstly carried out to study the mesh sensitivity of the FE model. For this purpose, a RC beam without FRP strengthening plate is analysed, so that FEM-P and FEM-B are the same in this case. The central deflections of the RC beam obtained from FE models with varying mesh densities (4007 elements, 7930 elements, and 9998 elements) are presented in Fig. 7.13. It can be seen that the computed central deflection converges to the experimental result when the number of elements is 7930. Thus, the finite element model with a mesh density of 7930 elements is used for the analysis of the FRP-strengthened RC beam.

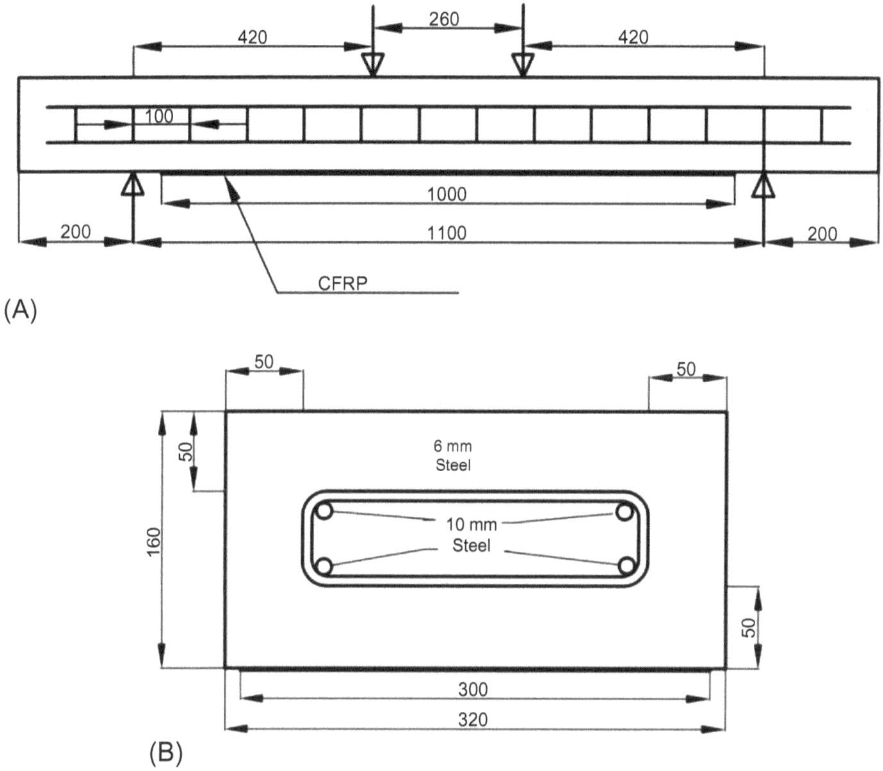

Fig. 7.12 A CFRP-strengthened RC beam tested by Arduini et al. [35] (dimension: millimetre). (A) Longitudinal section. (B) Cross section.

Table 7.1 Material properties of the CFRP-strengthened RC beam tested by Arduini et al. [35].

Material	Young's modulus (GPa)	Compressive strength (MPa)	Tensile strength (MPa)	Yield strength (MPa)	Poisson's ratio
Concrete	27	36	2.7	–	0.2
Steel	200	–	–	550	0.3
CFRP	235	–	3510	–	0.35
Epoxy	2.0	–	–	–	0.38

For the analysis of FRP-strengthened RC beams, a total of 10120 elements are employed, including 7679 SOLID65 elements, 251 LINK180 elements, 600 SHELL181 elements, 600 SOLID45 elements, and 990 COMBIN39 elements. The load–deflection relationship of the FRP-strengthened RC beam under static load is computed using both FEM-B and FEM-P. The numerical results obtained from both FEM-P and FEM-B are compared with the experimental test in Fig. 7.14. It can be seen clearly that the numerical results from the two finite element models generally agree well with the experiment. The

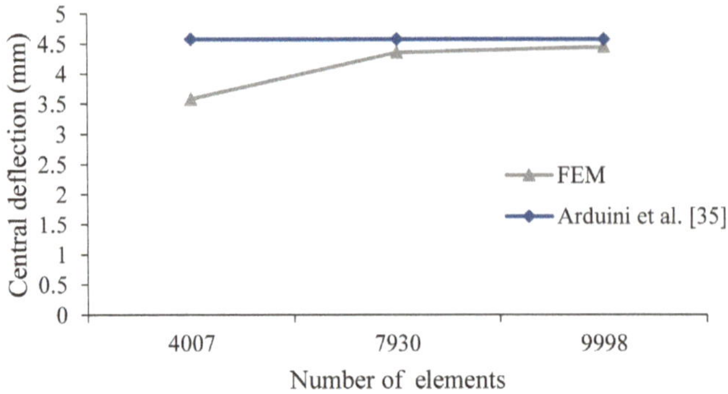

Fig. 7.13 Convergence test.
From P. Pathak, Y.X. Zhang, X. Teng, Nonlinear finite element analysis of FRP strengthened RC beams with bond-slip effect, Int. J. Comput. Method 14 (2) (2016) 17500321-19. With Permission from World Scientific Publishing Co., Inc.

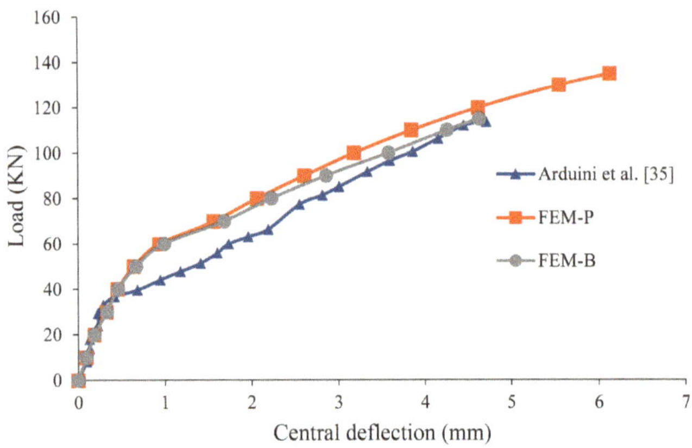

Fig. 7.14 Load versus central deflection of FRP-strengthened RC beam.
From P. Pathak, Y.X. Zhang, X. Teng, Nonlinear finite element analysis of FRP strengthened RC beams with bond-slip effect, Int. J. Comput. Method 14 (2) (2016) 17500321-19. With Permission from World Scientific Publishing Co., Inc.

two predicted curves are nearly identical before the load reaches 70 kN where no obvious bond–slip occurs. After that, bond–slip takes place, and the FEM-B model with bond–slip effect performs better in predicting the structural behaviour of FRP-strengthened RC beam than the FEM-P model with perfect bond assumption. At 115 kN, the central deflection obtained from the experiment is 4.8 mm, whilst the central deflections obtained from FEM-P and FEM-B are 4.12 and 4.63 mm, respectively.

Comparing the two finite element models, it can be seen from Fig. 7.14 that FEM-P provides a stiffer prediction of the flexural behaviour of the beam than the FEM-B model. After cracking, the FEM-P model overestimates the stiffness of the beam. This might be due to the fact that, when cracks occur, the perfect bond does not take the shear strain between concrete, adhesive, and CFRP into consideration. In addition, the effect of debonding failure on the flexural behaviour of the RC beam, which was observed in the experiments, is not considered in FEM-P. Thus, it can take the load further until another mode of failure occurs, such as FRP rupture, yielding of steel reinforcement, or concrete crushing.

7.7.2 CFRP-strengthened RC beam under cyclic load

In this section, a 5-metre-long CFRP-strengthened RC beam tested by Heffernan et al. [5] is analysed using FEM-PC and FEM-BC with the consideration of the effect of cyclic load on material performances. The CFRP-strengthened RC beam (5000 $\times$ 573.9 $\times$300 mm) is under a sinusoidal load for 630,000 cycles at a frequency of 3 Hz with the maximum and minimum amplitudes of 271 and 78 kN, respectively. The RC beam is reinforced with three longitudinal steel rebars of 25.2 mm diameter in tension side and two longitudinal rebars of 11.3 mm diameter in compression zone. The stirrups with 11.3 mm diameter are placed at a space of 250 mm centre-to-centre. The bottom side of the RC beam is externally bonded with 4,690-mm-long and 295-mm-wide FRP sheet. Details of the steel reinforcements and dimensions of the beam are illustrated in Fig. 7.15. The tensile strength of concrete is 4 MPa, and its compressive strength is 32.9 MPa. The Poisson's ratios of concrete and steel are 0.2 and 0.3, respectively. The elastic moduli of concrete, steel, adhesive, and CFRP are 27, 210, 2, and 325 GPa, respectively, and the tensile strengths of steel rebars and stirrups are 479 and 407 MPa, respectively.

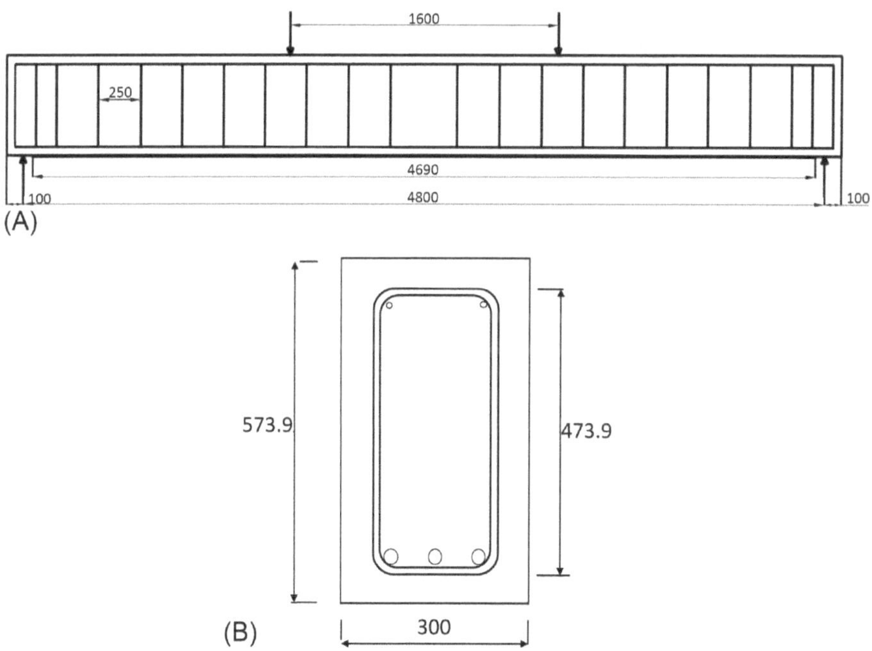

Fig. 7.15 CFRP-strengthened RC beam tested by Heffernan et al. [5] (dimensions: millimetre). (A) Longitudinal section. (B) Cross section.

Due to symmetry, only a quarter of the reinforced concrete beam is modelled. A convergence study is firstly carried out on a control RC beam without FRP strengthening plate. The central deflections of the control RC beam obtained from the FE models with 380 elements, 1778 elements, and 3420 elements are presented in Fig. 7.16. It can be seen that the computed central deflection converges to the experimental result (71.3 mm) when 1778 elements are employed. Thus, the finite element model with 1778 elements is used for the analysis of FRP-strengthened RC beam. In addition to the 1500 SOLID65 elements for concrete and 278 LINK180 elements for steel reinforcements, 141 SOLID45 elements, 141 SHELL181 elements, and 384 COMBIN39 elements are used to discretise adhesive, FRP, and interfaces between different components.

The central deflections of the RC beam against the number of cycles obtained from FEM-PC, FEM-BC, and experiment are compared in Fig. 7.17. It can be seen from Fig. 7.17 that the computed results agree with the experimental data well, especially the results from FEM-BC. Before 1500 loading cycles, the central deflection obtained from both numerical models coincides each other very well. After that, an obvious difference between the two numerical predictions can be seen, which shows the effect of the bond–slip in the finite element analysis. The maximum central deflection measured from the test is 14.6 mm after 3000 loading cycles. The predicted maximum central deflections after 3000 loading cycles are 14.5 mm from FEM-BC and 14.06 mm from FEM-PC. Both agree very well with the test data. The maximum central deflection of the FRP-strengthened RC beam obtained from the FEM-BC model after 630,000 loading cycles is 19.644 mm, which is close to the experimental result of 19.1 mm. However, the result from the FEM-PC model is only 16 mm, which might be due to the overestimation of stiffness and strength of FRP-strengthened RC beam, resulting from the absence of bond–slip between concrete, adhesive, and FRP interfaces in the numerical model. This further demonstrates the importance of the consideration of bond–slip effect in the numerical simulation of FRP-strengthened RC beams.

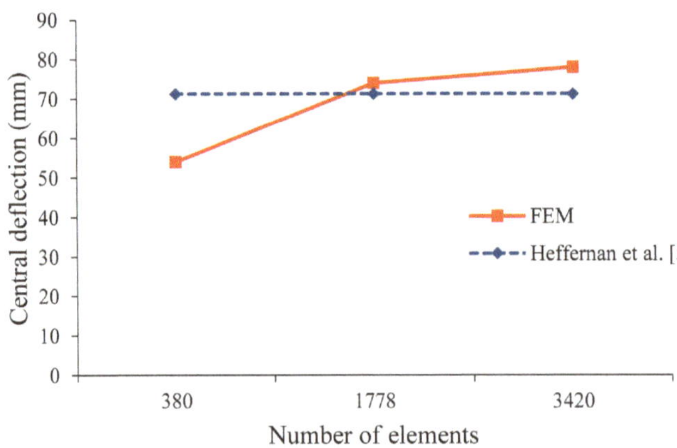

Fig. 7.16 Convergence test for control beam.
From P. Pathak, Y.X. Zhang, Nonlinear finite element analyses of FRP-strengthened steel-reinforced concrete beams under cyclic loading, Struct. Concr. 18 (2017) 929–937. With permission from John Wiley and Sons.

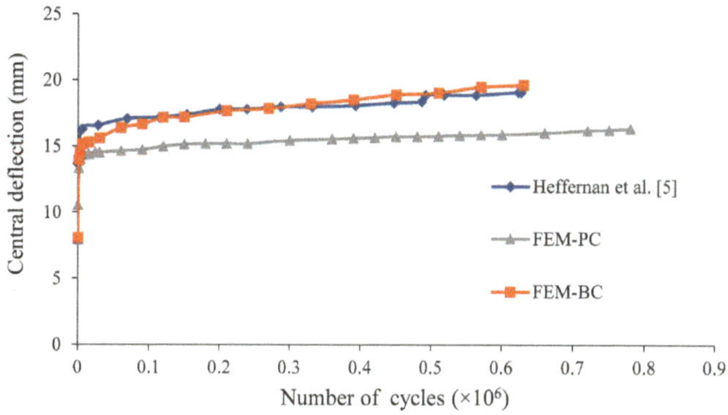

Fig. 7.17 Central deflection of CFRP-strengthened RC beam versus number of loading cycles. From P. Pathak, Y.X. Zhang, Numerical study of structural behavior of fiber-reinforced polymer-strengthened reinforced concrete beams with bond-slip effect under cyclic loading, Struct. Concr. 3 (2018) 1–11. With permission from John Wiley and Sons.

References

[1] M.F. Petrou, D. Parler, K.A. Harries, D.C. Rizos, Strengthening of reinforced concrete bridge decks using carbon fiber-reinforced polymer composite materials, J. Bridg. Eng. 13 (5) (2009) 455–467.

[2] S. Wang, H. Zhang, The study of fiber composite plates for strengthening reinforced bridges, in: A. Miravete (Ed.), Proceedings of Ninth International Conference on Composite Materials: Composites Properties and Applications, Madrid, Spain, 1993, pp. 224–231.

[3] R.A. Barnes, G.C. Mays, Fatigue performance of concrete beams strengthened with CFRP plates, J. Compos. Constr. 3 (2) (1999) 63–72.

[4] X. Teng, Y. Zhang, Nonlinear finite element analyses of FRP-strengthened concrete slabs under fixed-point cyclic loading, J. Compos. Constr. 19 (3) (2014) 04014057.

[5] P. Heffernan, M. Erki, Fatigue behavior of reinforced concrete beams strengthened with carbon fiber reinforced plastic laminates, J. Compos. Constr. 8 (2) (2004) 132–140.

[6] K.Y.M. Loo, S.J. Foster, S.T. Smith, FE modeling of CFRP-repaired RC beams subjected to fatigue loading, J. Compos. Constr. 16 (5) (2012) 572–580.

[7] J.O. Holmen, Fatigue of concrete by constant and variable amplitude loading, ACI Spec. Publ. 75 (1982) 71–110.

[8] Comité Euro-International du Béton (CEB-FIP), Model Code 1990. Paris, (1993).

[9] K.T. Chong, R.I. Gilbert, S.J. Foster, Modelling timedependent cracking in reinforced concrete using bond-slip interface elements, Comput. Concr. 1 (2) (2004) 151–168.

[10] E. Ferrier, D. Bigaud, J. Clement, P. Hamelin, Fatigue-loading effect on RC beams strengthened with externally bonded FRP, Constr. Build. Mater. 25 (2) (2011) 539–546.

[11] B. Täljsten, Strengthening of beams by plate bonding, J. Mater. Civ. Eng. 9 (4) (1997) 206–212.

[12] K. Aas-Jakobsen, Fatigue of Concrete Beams and Columns, Division of Concrete Structure Bulletin, Norwegian Institute of Technology, 1970.

[13] P. Balaguru, S. Shah, A method of predicting crack widths and deflections for fatigue loading, ACI Spec. Publ. 75 (1982) 153–176.

[14] J. Mander, F. Panthaki, A. Kasalanati, Low-cycle fatigue behavior of reinforcing steel, J. Mater. Civ. Eng. 6 (4) (1994) 453–468.

[15] P. Pathak, Y.X. Zhang, X. Teng, Nonlinear finite element analysis of FRP strengthened RC beams with bond-slip effect, Int. J. Comput. Method 14 (2) (2016) 17500321-19.
[16] P. Pathak, Y.X. Zhang, Nonlinear finite element analyses of FRP-strengthened steel-reinforced concrete beams under cyclic loading, Struct. Concr. 18 (2017) 929–937.
[17] P. Pathak, Y.X. Zhang, Numerical study of structural behavior of fiber-reinforced polymer-strengthened reinforced concrete beams with bond-slip effect under cyclic loading, Struct. Concr. 3 (2018) 1–11.
[18] M. Molina, J.J. Cruz, S. Oller, A.H. Barbat, L. Gil, Behaviour of the interface between concrete and FRP using serial/parallel mixing theory, Ing. Invest. 31 (3) (2011) 26–39.
[19] P. Fanning, Nonlinear models of reinforced and post-tensioned concrete beams, Electron. J. Struct. Eng. 2 (2001) 111–119.
[20] R. Shrestha, Behaviour of RC Beam-Column Connections Retrofitted With FRP Strips, PhD Thesis, School of Civil and Environmental Engineering, University of Technology Sydney, Sydney, Australia, 2009.
[21] N.P. Holmer, Parametric Study of the Bond Between Fiber Reinforced Polymers and Concrete Using Finite Element Analysis, Master Thesis, Faculty of the Graduate School, Marquette University, Milwaukee, Wisconsin, 2010.
[22] S. Popovics, A review of stress-strain relationships for concrete, ACI J. Proc. 67 (3) (1970) 243–248.
[23] E. Thorenfeldt, A. Tomaszewicz, J. Jensen, Mechanical properties of high-strength concrete and application in design, in: I. Holand (Ed.), Proceedings of the Symposium on Utilization of High Strength Concrete, Tapir, Trondheim, Norway, 1987, pp. 149–159.
[24] ANSYS, Inc, Release 15.0 Documentation for ANSYS, SAS IP Inc, USA, 2013.
[25] C.A. Coronado, M.M. Lopez, Sensitivity analysis of reinforced concrete beams strengthened with FRP laminates, Cem. Concr. Compos. 28 (1) (2006) 102–114.
[26] ACI Committee 318-99, Building Requirements for Structural Concrete, American Concrete Institute, Detroit, MI, 1999.
[27] K.J. William, E.D. Warnke, Constitutive model for the triaxial behavior of concrete, Proc. Int. Assoc. Bridge Struct. Eng. 19 (1975) 1–30.
[28] Y.T. Obaidat, Structural Retrofitting of Concrete Beams Using FRP, PhD Thesis, Department of Construction Sciences, Lund University, Sweden, 2011.
[29] A. Kinloch, Adhesion and Adhesives, Science and Technology, Chapman and Hall, New York, 1987.
[30] X. Lu, J. Teng, L. Ye, J. Jiang, Bond–slip models for FRP sheets/plates bonded to concrete, Eng. Struct. 27 (6) (2005) 920–937.
[31] K. Neale, U. Ebead, H.A. Baky, W. Elsayed, A. Godat, Analysis of the load-deformation behaviour and debonding for FRP-strengthened concrete structures, Adv. Struct. Eng. 9 (6) (2006) 751–763.
[32] C. Zanuy, L. Albajar, De la Fuente, Sectional analysis of concrete structures under fatigue loading, ACI Struct. J. 106 (5) (2009) 667–677.
[33] K. Loo, Fatigue Behaviour of CFRP-Repaired Corroded RC Beams, PhD Thesis, School of Civil and Environmental Engineering, University of New South Wales, Sydney, Australia, 2010.
[34] U. Muralidharan, S. Manson, A modified universal slopes equation for estimation of fatigue characteristics of metals, J. Eng. Mater. Technol. 110 (1) (1988) 55–58.
[35] M. Arduini, A. Nanni, Behavior of precracked RC beams strengthened with carbon FRP sheets, J. Compos. Constr. 1 (2) (1997) 63–70.

Appendix A: List of notations

Notation	Description
A	Cross-sectional area
$A_{c,eff}$	Concrete area involved in the stiffening
a_{ij}	Unknown coefficients
$[B_0]$	Linear part of the strain matrix
$[B_b]$	Curvature–displacement matrix
$[B_{bar}]$	Axial strain–displacement matrix of the reinforcing layers
$[B_l]$	Nonlinear part of the strain matrix
$[B_s]$	Shear strain–displacement matrix
b	Width of the beam element
b_{ij}	Unknown coefficients
b_i	Width of the ith layer of the beam element
c_c	Specific heat
D_{bb}	Bending stiffness
D_{ss}	Transverse shear stiffness
$[D_l]$	Tangential modulus matrix
d	Rebar diameter
E	Modulus of elasticity
E_{bond}	Modulus of bond
E_T	Elastic modulus at temperature T
$\{F\}$	Vector of nodal forces
$\{F_T^{(e)}\}$	Thermal load vector
$\{f^{(e)}\}$	Equivalent nodal load vector
f'_c	Concrete compressive strength
f_{cT}	Concrete compressive strength at temperature T
f_f	Ultimate tensile strength of FRP
f_{fT}	Ultimate tensile strength of FRP rebars at temperature T
f_t	Tensile strength of concrete
f_{tT}	Concrete tensile strength at temperature T
f_u	Ultimate tensile strength of steel
f_y	Yield strength of steel
f_{yT}	Yield strength of steel at temperature T
G	Shear modulus
G_c	Uncracked concrete shear modulus
G_{cT}	Shear modulus at temperature T
G'_c	Concrete shear modulus after cracking
G'_{cT}	Cracked shear modulus at temperature T
G_f	Interfacial fracture energy
h	Thickness of the beam element

Continued

h_T	Convection coefficient
I	Moment of inertia
$[K]$	Tangential stiffness matrix
$[K_0^{(e)}]$	Linear stiffness matrix
$[K_l^{(e)}]$	Large displacement matrix
$[K_\sigma^{(e)}]$	Initial stress matrix
$[K_T^{(e)}]$	Element conduction matrix
k_c	Reduction factor of the concrete compressive strength at temperature T
k_{cc}	Concrete thermal conductivity
L	Length of the beam element
M	Bending moment
$[M^{(e)}]$	Element mass matrix
N	Axial force
$[N]$	Shape function matrix
p	Perimeter of the rebar
Q	Internal heat source
q_h	Heat flow
$\{q\}$	Vector of nodal displacement
R	Radius of curvature at a point on the neutral axis of the beam
$\{R\}$	Vector of externally applied nodal loads
s	Bond–slip at interface
s_0	Local bond–slip at τ_{max}
s_f	Local bond–slip at $\tau=0$
T	Temperature
t	Equivalent thicknesses of smeared reinforcing layers
t_{ci}, t_{sj}	Thicknesses of the ith concrete layer and jth reinforcement layer
$\{t^{(e)}\}$	Element nodal temperature vector
$\left\{\dot{t}^{(e)}\right\}$	Temperature derivative with respect to time
u	Longitudinal displacement
u_s	Axial displacement of the reinforcing layer with bond–slip
w	Transverse displacement of the beam
w_0	Transverse deflection in the midplane of the beam
w_1, w_2	Nodal displacements
Y	Distance from a point along the z direction to the neutral axis
y_n	Position of the neutral axis
α	Ratio of the beam bending stiffness to the shear stiffness
β	Tension-stiffening factor
β_T	Tension-stiffening factor at temperature T
β_t	Shear transfer coefficient
β_w	Width ratio factor
δ	Shear correction factor
ε_{bond}	Strain due to slip
ε_0	Concrete compressive strain at compressive strength
ε_c	Strain of the concrete layer
ε_{cu}	The ultimate compressive strain of the concrete
ε_{cuT}	Strain corresponding to f_{cT}
ε_x	Longitudinal strain

ε_b	Bending strain
ε_s	Strain of the reinforcing layer
ε_T	Strain of steel at temperature T
ε_{th}	Thermal expansion
ε_{me}	Mechanical strain
$\{\varepsilon_0\}$	Linear strain vector
$\{\varepsilon_l\}$	Nonlinear strain vector
ϕ	Slope du/dx in the midplane of the beam
γ	Transverse shear strain
κ	Bond modulus
λ	Tolerance of the displacements
θ	Rotation of a transverse normal about the y-axis
θ_1, θ_2	Nodal rotations
ρ_b	Balanced reinforcement ratio
ρ_c	Concrete mass density
ρ_{eff}	Effective reinforcement ratio
ρ_f	Reinforcement ratio of the FRP rebars
σ_{cu}	Ultimate compressive stress
σ_c	Concrete compressive stress
σ_t	Concrete tensile stress
σ_T	Stress of steel at temperature T
$\{\sigma\}$	Internal stress vector of the element
τ	Bond stress at interface
τ_{max}	Maximum local bond stress
υ	Poisson's ratio
ξ	Unknown function of position x
$\{\psi^{(e)}\}$	Unbalanced force of the beam element

Appendix B: Gaussian integration

There are various Gaussian rules, where Gauss–Legendre rules are the most important and widely used in the integration. The Legendre polynomials form an orthogonal family, and other orthogonal polynomials can be used to handle awkward integrals. The Gauss–Legendre approach requires the integration to be carried out over the range $[-1 \quad 1]$ [1]. The Gaussian points are the roots of Legendre polynomials, which take the form of

$$P_n(x) = \frac{1}{2^n n!} \frac{d^n}{dx^n} \left(x^2 - 1\right)^n = 0 \tag{B.1}$$

where n is the number of Gaussian points and x is the location at the x-axis. The weights and locations for the first six cases are shown in Table B.1.

Table B.1 Locations and weights for normalised Gauss–Legendre rules in 1-D problem [1].

Number of Gaussian points	Locations	Weights
1	0	2
2	±0.5773502692	1
3	±0.7745966692	0.5555555556
	0	0.8888888889
4	±0.8611363116	0.3478548451
	±0.3399810436	0.6521451549
5	±0.9061798459	0.2369268851
	±0.5384693101	0.4786286705
	0	0.5688888889
6	±0.9324695142	0.1713244924
	±0.6612093865	0.3667615730
	±0.2386191861	0.4679139346

Reference

[1] A.C. Bajpai, L.R. Mustoe, D. Walker, Advanced Engineering Mathematics, John Wiley and Sons Ltd., Great Britain, 1977.

Appendix C: Temperature-dependent material properties of concrete

C.1 Thermal elongation of concrete

The thermal elongation ε_{th} of concrete with the temperature T (°C) may be determined by the following equations [1].
Concrete with siliceous aggregates:
For $20 < T \leq 700$

$$\varepsilon_{th} = -1.8 \times 10^{-4} + 9 \times 10^{-6} T + 2.3 \times 10^{-11} T^3 \tag{C.1a}$$

For $700 < T \leq 1200$

$$\varepsilon_{th} = 14 \times 10^{-3} \tag{C.1b}$$

Concrete with calcareous aggregates:
For $20 < T \leq 805$

$$\varepsilon_{th} = -1.2 \times 10^{-4} + 6 \times 10^{-6} T + 1.4 \times 10^{-11} T^3 \tag{C.2a}$$

For $805 < T \leq 1200$

$$\varepsilon_{th} = 12 \times 10^{-3} \tag{C.2b}$$

C.2 Thermal conductivity, mass density, and specific heat of concrete

For normal weight concrete, the thermal conductivity k_{cc} (W/m °C) may be determined from [2]

$$k_{cc} = 2 - 0.2451(T/100) + 0.0107(T/100)^2 \text{ for } 20 \leq T \leq 1200 \tag{C.3}$$

The specific heat c_c (J/kg °C) of normal weight dry, siliceous or calcareous concrete may be determined from [2]

$$c_c = 900 \text{ for } 20 \le T \le 100 \quad \text{(C.4a)}$$

$$c_c = 900 + (T - 100) \text{ for } 100 < T \le 200 \quad \text{(C.4b)}$$

$$c_c = 1000 + (\mathrm{T} - 200)/2 \text{ for } 200 < T \le 400 \quad \text{(C.4c)}$$

$$c_c = 1100 \text{ for } 400 < T \le 1200 \quad \text{(C.4d)}$$

For static loads, the mass density of concrete ρ_c (kg/m^3) may be considered to be independent of the concrete temperature. For calculation of the thermal response, the variation of ρ_c as a function of temperature may be considered as

$$\rho_c = 2354 - 23.47(T/100) \quad \text{(C.5)}$$

C.3 Values for the main parameters of concrete constitutive relationship at elevated temperatures

The strength and deformation properties of uniaxial stressed concrete at elevated temperatures are characterised by a set of stress–strain relationships with a shape as specified in Fig. C.1 [1,2].

For a given concrete temperature T (°C), the stress–strain curve is defined by parameters f_{cT}, ε_{cuT}, and ε_{ceT}, where f_{cT} is determined by the reduction factor k_c ($f_{cT} = k_c f_c$). Values for k_c, ε_{cuT}, and ε_{ceT} are given as follows.

The reduction factor k_c for concrete with siliceous aggregates [2]:

$k_c = 1$ for $T \le 100$
$k_c = 1.05 - 0.0005T$ for $100 < T \le 200$
$k_c = 1.15 - 0.001T$ for $200 < T \le 400$
$k_c = 1.35 - 0.0015T$ for $400 < T \le 800$
$k_c = 0.71 - 0.0007T$ for $800 < T \le 900$
$k_c = 0.44 - 0.0004T$ for $900 < T \le 1000$

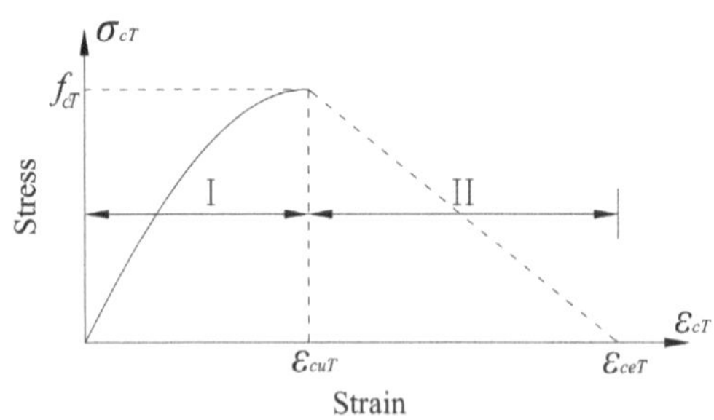

Fig. C.1 Stress–strain relationship of concrete under compression at elevated temperatures [1,2].

$k_c = 0.34 - 0.0003T$ for $1000 < T \leq 1100$
$k_c = 0.12 - 0.0001T$ for $1100 < T \leq 1200$
$k_c = 0$ for $1200 < T$

The reduction factor k_c for concrete with calcareous aggregates [1]:

$k_c = 1$ for $T = 20$
$k_c = 1.0075 - 0.000375T$ for $20 < T \leq 100$
$k_c = 1 - 0.0003T$ for $100 < T \leq 300$
$k_c = 1.09 - 0.0006T$ for $300 < T \leq 400$
$k_c = 1.29 - 0.0011T$ for $400 < T \leq 500$
$k_c = 1.44 - 0.0014T$ for $500 < T \leq 600$
$k_c = 1.62 - 0.0017T$ for $600 < T \leq 700$
$k_c = 1.55 - 0.0016T$ for $700 < T \leq 800$
$k_c = 1.23 - 0.0012T$ for $800 < T \leq 900$
$k_c = 0.96 - 0.0009T$ for $900 < T \leq 1000$
$k_c = 0.46 - 0.0004T$ for $1000 < T \leq 1100$
$k_c = 0$ for $1100 < T$

Parameters ε_{cuT} and ε_{ceT} defining the range of the descending branch for the stress–strain relationship of concrete at elevated temperatures are determined as [2]

$\varepsilon_{cuT} = 0.0025$ for $T = 20$
$\varepsilon_{cuT} = 0.0025 + 0.00001875(T - 20)$ for $20 < T \leq 100$
$\varepsilon_{cuT} = 0.004 + 0.000015(T - 100)$ for $100 < T \leq 300$
$\varepsilon_{cuT} = 0.007 + 0.00003(T - 300)$ for $300 < T \leq 400$
$\varepsilon_{cuT} = 0.01 + 0.00005(T - 400)$ for $400 < T \leq 500$
$\varepsilon_{cuT} = 0.015 + 0.0001(T - 500)$ for $500 < T \leq 600$
$\varepsilon_{cuT} = 0.025$ for $600 < T \leq 1100$

and

$\varepsilon_{ceT} = 0.02$ for $T = 20$
$\varepsilon_{ceT} = 0.02 + 0.00003125(T - 20)$ for $20 < T \leq 100$
$\varepsilon_{ceT} = 0.0225 + 0.000025(T - 100)$ for $100 < T \leq 1100$

References

[1] European Committee for Standardization, in: M. Keynes (Ed.), Eurocode 2: Design of Concrete Structures. Part 1–2. General Rules-Structural Fire Design (DD ENV 1992-1-2), British Standards Institution, 1995.
[2] European Committee for Standardization, Eurocode 4: Design of Composite Steel and Concrete Structures. Part 1–2. General Rules-Structural Fire Design (BS EN 1994-1-2), 2005.

Appendix D: Temperature-dependent material properties of steel

D.1 Thermal elongation of steel

The change of thermal elongation ε_{th} of reinforcing steel with the temperature T (°C) may be adopted according to Eqs. (D.1a)–(D.1c) [1].
For $20 < T \leq 750$

$$\varepsilon_{th} = -2.416 \times 10^{-4} + 1.2 \times 10^{-5} T + 0.4 \times 10^{-8} T^2 \tag{D.1a}$$

For $750 < T \leq 860$

$$\varepsilon_{th} = 11 \times 10^{-3} \tag{D.1b}$$

For $T \geq 860$

$$\varepsilon_{th} = -6.2 \times 10^{-3} + 2 \times 10^{-5} T \tag{D.1c}$$

D.2 Values for the main parameters of steel constitutive relationship at elevated temperatures

The strength and deformation properties of steel at elevated temperatures are characterised by a set of stress–strain relationships with a shape as specified in Fig. D.1 [2].

For a given steel temperature T (°C), the stress–strain curve is defined by parameters f_{pT}, f_{yT}, and f_{uT}, which are determined by reduction factors k_{pT}, k_{yT}, and k_{uT}, respectively. The modulus of elasticity of steel E_{sT} also decreases with the increase of temperature and is adjusted by a reduction factor k_{ET}. Values of these reduction factors are given as follows.

Modulus of elasticity of steel: $E_{sT} = k_{ET} E_s$

$k_{ET} = 1$ for $T \leq 100$
$k_{ET} = 1.1 - 0.001T$ for $100 < T \leq 500$

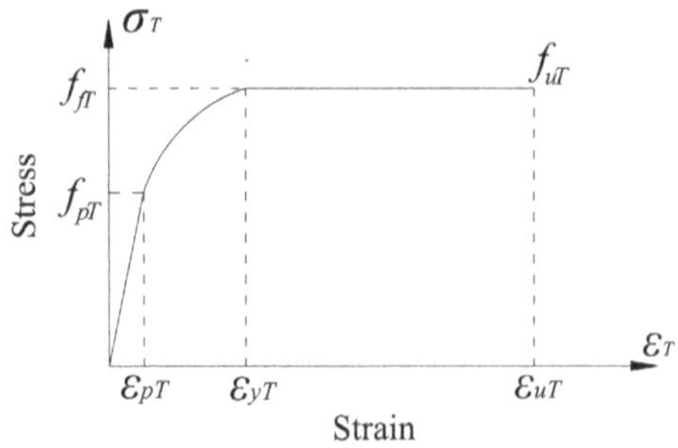

Fig. D.1 Stress–strain relationship of steel at elevated temperatures [2].

$k_{ET}=2.05-0.0029T$ for $500<T\leq 600$
$k_{ET}=1.39-0.0018T$ for $600<T\leq 700$
$k_{ET}=0.41-0.0004T$ for $700<T\leq 800$
$k_{ET}=0.27-0.000225T$ for $800<T\leq 1200$

Proportional limit: $f_{pT}=k_{pT}f_y$

$k_{pT}=1$ for $T\leq 100$
$k_{pT}=1.193-0.00193T$ for $100<T\leq 200$
$k_{pT}=1.195-0.00194T$ for $200<T\leq 300$
$k_{pT}=1.192-0.00193T$ for $300<T\leq 400$
$k_{pT}=0.66-0.0006T$ for $400<T\leq 500$
$k_{pT}=1.26-0.0018T$ for $500<T\leq 600$
$k_{pT}=0.81-0.00105T$ for $600<T\leq 700$
$k_{pT}=0.25-0.00025T$ for $700<T\leq 800$
$k_{pT}=0.15-0.000125T$ for $800<T\leq 1200$

Yield strength: $f_{yT}=k_{yT}f_y$

$k_{yT}=1$ for $T\leq 400$
$k_{yT}=1.88-0.0022T$ for $400<T\leq 500$
$k_{yT}=2.33-0.0031T$ for $500<T\leq 600$
$k_{yT}=1.91-0.0024T$ for $600<T\leq 700$
$k_{yT}=1.07-0.0012T$ for $700<T\leq 800$
$k_{yT}=0.51-0.0005T$ for $800<T\leq 900$
$k_{yT}=0.24-0.0002T$ for $900<T\leq 1200$

Ultimate strength: $f_{uT}=k_{uT}f_y$

$k_{uT}=1.25$ for $T\leq 300$
$k_{uT}=2-0.0025T$ for $300<T\leq 400$
$k_{uT}=1.88-0.0022T$ for $400<T\leq 500$
$k_{uT}=2.33-0.0031T$ for $500<T\leq 600$
$k_{uT}=1.91-0.0024T$ for $600<T\leq 700$
$k_{uT}=1.07-0.0012T$ for $700<T\leq 800$
$k_{uT}=0.51-0.0005T$ for $800<T\leq 900$
$k_{uT}=0.24-0.0002T$ for $900<T\leq 1200$

References

[1] European Committee for Standardization, in: M. Keynes (Ed.), Eurocode 2: Design of Concrete Structures. Part 1–2. General Rules-Structural Fire Design (DD ENV 1992-1-2), British Standards Institution, 1995.

[2] European Committee for Standardization, Eurocode 4: Design of Composite Steel and Concrete Structures. Part 1–2. General Rules-Structural Fire Design (BS EN 1994-1-2), 2005.

Appendix E: Temperature-dependent material properties of FRP

The strength and deformation properties of FRP at elevated temperatures are characterised by a set of stress–strain relationships with a shape assumed to be linear elastic until fracture as specified in Fig. E.1.

For a given FRP temperature T (°C), the stress–strain curve is defined by parameters f_{fT} and E_{fT}, which are determined using reduction factors k_{fT} ($f_{fT} = k_{fT} f_f$) and k_{fET} ($E_{fT} = k_{fET} E_f$), respectively. The values of these reduction factors are given as follows:

For GFRP rebars [1]

$k_{fT} = 1 - 0.0025T$ for $0 \leq T \leq 400$
$k_{fT} = 0$ for $400 \leq T$

For AFRP rebars [1]

$k_{fT} = 1$ for $0 \leq T \leq 100$
$k_{fT} = 1.333 - 0.00333T$ for $100 \leq T \leq 400$
$k_{fT} = 0$ for $400 \leq T$

For CFRP rebars [2]

$k_{fT} = 1$ for $0 \leq T \leq 200$
$k_{fT} = 1.381 - 0.0019T$ for $200 \leq T \leq 725$
$k_{fT} = 0$ for $725 \leq T$

For GFRP and AFRP rebars [1]

$k_{fET} = 1$ for $0 \leq T \leq 100$
$k_{fET} = 1.25 - 0.0025T$ for $100 \leq T \leq 300$
$k_{fET} = 2 - 0.005T$ for $300 \leq T \leq 400$
$k_{fET} = 0$ for $400 \leq T$

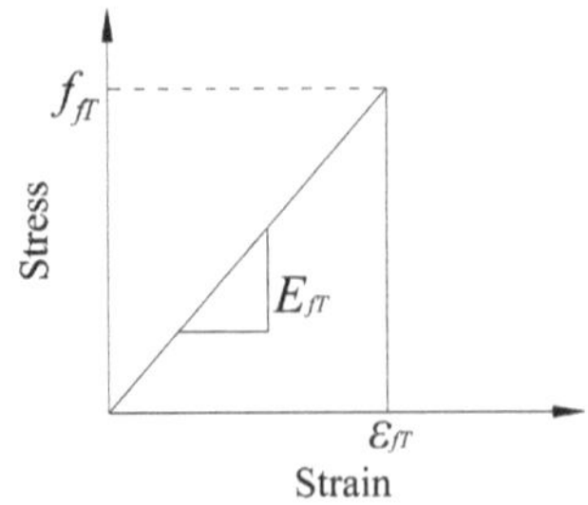

Fig. E.1 Stress–strain relationship of FRP at elevated temperatures.

For CFRP rebars [2]

$$k_{fET} = 1 \text{ for } 0 \leq T \leq 200$$
$$k_{fET} = 1.28 - 0.0014T \text{ for } 200 \leq T \leq 450$$
$$k_{fET} = 1.625 - 0.00217T \text{ for } 450 \leq T \leq 750$$
$$k_{fET} = 0 \text{ for } 750 \leq T$$

References

[1] M. Saafi, Effect of fire on FRP reinforced concrete members, Compos. Struct. 58 (1) (2002) 11–20.

[2] M.M. Rafi, A. Nadjai, F. Ali, Finite element modeling of carbon fiber-reinforced polymer reinforced concrete beams under elevated temperatures, ACI Struct. J. 105 (6) (2008) 701–710.

Appendix F: Finite element code for composite beam element: Linear analysis

This appendix provides the principal finite element code for linear analysis of homogeneous and composite beams using the 1-D composite beam element introduced in this book. The program is written in Fortran 90.

```
{Module definition for common variables is omitted herein}
PROGRAM COMPOSITE_BEAM
!--------------------------------------------------------------------------------
! This program is used for linear analysis of
! homogeneous and composite beams.
!--------------------------------------------------------------------------------
 USE MODULE_NAME
 IMPLICIT NONE
 INTEGER :: STIFF_SIZE, COUNT_NUM, COUNT_j, n, m, MAX, i, j

 CALL INPUT_DATA
 !-------------------------------------------------------------------------------
 ! The subroutine INPUT_DATA reads in the basic
 ! information of the model, including
 ! dimension of beam, number of material, locations
 ! and thicknesses of different materials, basic
 ! material properties, element size, number of
 ! layers, support conditions, and loading conditions.
 !-------------------------------------------------------------------------------

 ! Call subroutine to generate element stiffness
 CALL ELE_STIFF_LINEAR

 {Allocate array dimensions for nodal load, stiffness, displacement,
strain, and stress.}

 DIS_Global_Step = 0
 ROTA_Global_Step = 0
 NONDIS_Global_Step = 0
 NONROTA_Global_Step = 0
 Displacement_Vec = 0
```

```
DO STEP = 1, NUM_LOAD_STEP

 ! Call subroutine to find nodal displacements
 CALL Global_Displacement_Linear

 DIS_Global_Step = DIS_Global_Step + DIS_Global
 ROTA_Global_Step = ROTA_Global_Step + ROTA_Global

 WRITE(1,*) 'Global vertical displacements of the', STEP,'step are:'
 DO COUNT_j = 1, Num_Node_Stru
  WRITE(1,*) 'DIS_Global_Step(', COUNT_j, ') =', DIS_Global_Step
  (COUNT_j,1)
 END DO

 WRITE(1,*) 'Global rotations of the', STEP,'step are:'
 DO COUNT_j = 1, Num_Node_Stru
  WRITE(1,*) 'ROTA_Global_Step(', COUNT_j, ') =', ROTA_Global_Step
(COUNT_j,1)
 END DO

 {Deallocate arrays for stiffness and displacement.}

END DO

END PROGRAM COMPOSITE_BEAM


SUBROUTINE Global_Displacement_Linear
!------------------------------------------------------------------------------
! The subroutine Global_Displacement_Linear finds
! linear global nodal displacements.
!------------------------------------------------------------------------------
 USE MODULE_NAME
 IMPLICIT NONE
 INTEGER :: COUNT_NUM, COUNT_i, COUNT_j, COUNT_AUG1, M, CC

 ! Call subroutine to apply boundary conditions
 CALL Global_Boundary_Linear

 M = Num_Unknown+1

 ALLOCATE(AUG(Num_Unknown,M), AUG_1(DIMENSION_GLOBAL,M))
 ALLOCATE(Displacement_Vec(Num_Unknown,1))

 AUG = 0
 AUG_1 = 0

 ! Find zero columns
 C_0 = DIMENSION_GLOBAL-Num_Unknown
 ALLOCATE(COLUMN_ZERO(C_0))
```

```
SKIP = 0
DO COUNT_j = 1,DIMENSION_GLOBAL
    SUM_COLUMN = 0

 DO COUNT_i = 1,DIMENSION_GLOBAL
  IF (K_Global_Bound(COUNT_i,COUNT_j) /= 0) THEN
   SUM_COLUMN = SUM_COLUMN + 1
  END IF
 END DO

 IF (SUM_COLUMN == 0) THEN
  SKIP = SKIP+1
  COLUMN_ZERO(SKIP) = COUNT_j
 END IF

END DO

DO COUNT_i = 1, DIMENSION_GLOBAL
 SKIP = 0
 DO COUNT_j = 1, DIMENSION_GLOBAL
  NONZERO = 0

  DO COUNT_NUM = 1, C_0
   IF (COUNT_j == COLUMN_ZERO(COUNT_NUM)) THEN
    NONZERO = NONZERO + 1
   END IF
  END DO

  IF (NONZERO == 0) THEN
   SKIP = SKIP+1
   AUG_1(COUNT_i,SKIP) = K_Global_Bound(COUNT_i,COUNT_j)
  END IF

 END DO
END DO

! Find zero rows
R_0 = DIMENSION_GLOBAL-Num_Unknown
ALLOCATE(ROW_ZERO(R_0))

SKIP = 0
DO COUNT_i = 1, DIMENSION_GLOBAL
    SUM_COLUMN = 0

 DO COUNT_j = 1, DIMENSION_GLOBAL
  IF (K_Global_Bound(COUNT_i,COUNT_j) /= 0) THEN
```

```
   SUM_COLUMN = SUM_COLUMN + 1
  END IF
 END DO

 IF (SUM_COLUMN == 0) THEN
  SKIP = SKIP+1
  ROW_ZERO(SKIP) = COUNT_i
 END IF

END DO

DO COUNT_j = 1, Num_Unknown
 SKIP = 0

 DO COUNT_i = 1, DIMENSION_GLOBAL
  NONZERO = 0

  DO COUNT_NUM = 1, R_0
   IF (COUNT_i == ROW_ZERO(COUNT_NUM)) THEN
    NONZERO = NONZERO + 1
   END IF
  END DO

  IF (NONZERO == 0) THEN
   SKIP = SKIP+1
   AUG(SKIP,COUNT_j) = AUG_1(COUNT_i,COUNT_j)
  END IF

 END DO
END DO

ALLOCATE(Nodal_Load(DIMENSION_GLOBAL,1))

Nodal_Load = LOAD_COE*(Con_Load + Dis_Load)

SKIP = 0
DO COUNT_i = 1,DIMENSION_GLOBAL
 Nodal_Load(COUNT_i,1)=Nodal_Load(COUNT_i,1) * SUPPORT(COUNT_i,1)
     NONZERO = 0

 DO COUNT_NUM = 1, R_0
  IF (COUNT_i == ROW_ZERO(COUNT_NUM)) THEN
   NONZERO = NONZERO + 1
  END IF
 END DO

 IF (NONZERO == 0) THEN
  SKIP = SKIP+1
```

```
   AUG(SKIP,M) = Nodal_Load(COUNT_i,1)
  END IF

 END DO

 ! Call subroutines to find displacements
 CALL GAUSS_ELIMINATION

 IF (SOLVBL) THEN
  CALL BACK(Displacement_Vec)
 ELSE
  WRITE(*,*) 'NO SOLUTION EXISTS'
 END IF

 ALLOCATE(DIS_Global(Num_Node_Stru,1), ROTA_Global(Num_Node_Stru,1))

 CC = 0
 DO COUNT_i = 1, DIMENSION_GLOBAL
  IF (SUPPORT(COUNT_i,1) == 0) THEN
   Displacement_Global(COUNT_i,1) = 0
  ELSE
   CC = CC + 1
   Displacement_Global(COUNT_i,1) = Displacement_Vec(CC,1)
  END IF
 END DO

 DO COUNT_j = 1, Num_Node_Stru
  DIS_Global(COUNT_j,1) = Displacement_Global(2*COUNT_j-1,1)
  ROTA_Global(COUNT_j,1) = Displacement_Global(2*COUNT_j,1)
 END DO

 DEALLOCATE(Nodal_Load)

END SUBROUTINE Global_Displacement_Linear


SUBROUTINE Global_Boundary_Linear
!-------------------------------------------------------------------------------
! The subroutine Global_Boundary_Linear applies
! boundary conditions onto global stiffness matrix.
!-------------------------------------------------------------------------------
 USE MODULE_NAME
 IMPLICIT NONE
 INTEGER :: i, j

 ! Call subroutine to generate global stiffness matrix
 CALL Global_Stiff_Matrix_Linear

 ! Allocate array dimensions for global stiffness matrix
 ALLOCATE(Kb_Global_Bound(DIMENSION_GLOBAL,DIMENSION_GLOBAL))
 ALLOCATE(Ks_Global_Bound(DIMENSION_GLOBAL,DIMENSION_GLOBAL))
```

```
 DO i = 1,DIMENSION_GLOBAL
  DO j = 1,DIMENSION_GLOBAL
   Kb_Global_Bound(i,j) = Kb_Global(i,j) * SUPPORT(i,1)
   Ks_Global_Bound(i,j) = Ks_Global(i,j) * SUPPORT(i,1)
  END DO
 END DO

 DO j = 1,DIMENSION_GLOBAL
  DO i = 1,DIMENSION_GLOBAL
   Kb_Global_Bound(i,j) = Kb_Global_Bound(i,j) * SUPPORT(j,1)
   Ks_Global_Bound(i,j) = Ks_Global_Bound(i,j) * SUPPORT(j,1)
  END DO
 END DO

 ALLOCATE(K_Global_Bound(DIMENSION_GLOBAL,DIMENSION_GLOBAL))
 DO i = 1,DIMENSION_GLOBAL
  DO j = 1,DIMENSION_GLOBAL
   K_Global_Bound(i,j)=Kb_Global_Bound(i,j)+Ks_Global_Bound(i,j)
  END DO
 END DO

END SUBROUTINE Global_Boundary_Linear


SUBROUTINE Global_Stiff_Matrix_Linear
!--------------------------------------------------------------------------------
! The subroutine Global_Stiff_Matrix_Linear
! generates global stiffness matrix.
!--------------------------------------------------------------------------------
 USE MODULE_NAME
 IMPLICIT NONE
 INTEGER :: COUNT_DIMEN, COUNT_i, COUNT_j, locate

 ! Call subroutine to generate element stiffness matrix
 CALL STIFF_MATRIX_LINEAR

 ! Allocate array dimensions for global stiffness matrix
 ALLOCATE(Kb_Global(DIMENSION_GLOBAL,DIMENSION_GLOBAL))
 ALLOCATE(Ks_Global(DIMENSION_GLOBAL,DIMENSION_GLOBAL))
 ALLOCATE(K_Global(DIMENSION_GLOBAL,DIMENSION_GLOBAL))

 locate = 0
 Kb_Global = 0
 Ks_Global = 0

 DO COUNT_DIMEN = 1, Num_Element_Stru
  DO COUNT_i = 1, 4
```

```
    DO COUNT_j = 1, 4
     Kb_Global(COUNT_i+locate,COUNT_j+locate) = Kb_Global(COUNT_i+
locate,COUNT_j+locate) + Kb(COUNT_i,COUNT_j)
     Ks_Global(COUNT_i+locate,COUNT_j+locate) = Ks_Global(COUNT_i+
locate,COUNT_j+locate) + Ks(COUNT_i,COUNT_j)
    END DO
   END DO

   locate = locate + 2

  END DO

  DO COUNT_i = 1, DIMENSION_GLOBAL
   DO COUNT_j = 1, DIMENSION_GLOBAL
    K_Global(COUNT_i,COUNT_j) = Kb_Global(COUNT_i,COUNT_j) + Ks_Global
(COUNT_i,COUNT_j)
   END DO
  END DO

END SUBROUTINE Global_Stiff_Matrix_Linear


SUBROUTINE STIFF_MATRIX_LINEAR
!---------------------------------------------------------------------------------
! The subroutine STIFF_MATRIX_LINEAR generates
! bending and shear stiffness matrix for element.
!---------------------------------------------------------------------------------
 USE MODULE_NAME
 IMPLICIT NONE
 REAL, ALLOCATABLE, DIMENSION(:) :: u, w
 INTEGER :: i,j,COUNT_Gauss

 ! Call subroutine to select gauss integral constants
 CALL Gauss_Beam(u,w)

 Kb = 0
 Ks = 0

 DO i = 1,4
  DO j = 1,4
   DO COUNT_Gauss = 1, Num_Gauss_Point
    x = 0.5*L*u(COUNT_Gauss) + 0.5*L
    CALL Shape_Function_Linear
    Kb1 = MATMUL(TRANSPOSE(Shape_Bb),Shape_Bb)* Bending_Stiff
    Kb(i,j) = Kb(i,j) + 0.5*L*w(COUNT_Gauss)*Kb1(i,j)
    Ks1 = MATMUL(TRANSPOSE(Shape_Bs),Shape_Bs)* Shear_Stiff
    Ks(i,j) = Ks(i,j) + 0.5*L*w(COUNT_Gauss)*Ks1(i,j)
   END DO
  END DO
 END DO
```

```
 K0 = Kb + Ks

END SUBROUTINE STIFF_MATRIX_LINEAR


SUBROUTINE ELE_STIFF_LINEAR
!------------------------------------------------------------------------------
! The subroutine ELE_STIFF_LINEAR generates
! element stiffness for linear analysis.
!------------------------------------------------------------------------------
 USE MODULE_NAME
 IMPLICIT NONE
 INTEGER :: COUNT_MAT, MAX, i, j
 REAL :: FE, FEy

 ! Allocate stiffness of composite beam element
 ALLOCATE(Dbb(Num_Material_Ele),Dss(Num_Material_Ele))
 ALLOCATE(Dxl_LIN(Num_Material_Ele),Dxb_LIN(Num_Material_Ele))

 MAX = Num_Layer_Material(1)
 IF (Num_Material_Ele > 1) THEN
  DO i = 2, Num_Material_Ele
   IF (Num_Layer_Material(i) >= MAX) THEN
    MAX = Num_Layer_Material(i)
   END IF
  END DO
 END IF

 ! Allocate coordinates of materials
 ALLOCATE(Material_Top(Num_Material_Ele,MAX),Material_Bot
 (Num_Material_Ele,MAX))

 ! For homogeneous beam
 DO ilayer = 1, Num_Layer_Material(1)
  Material_Bot(1,ilayer)=START_CO(1)+(ilayer-1)*Thickness/
  Num_Layer_Material(1)
  Material_Top(1,ilayer) = START_CO(1) + ilayer*Thickness/
  Num_Layer_Material(1)
  WRITE(10,*) 'Material_Bot(1,',ilayer,')=', Material_Bot(1,ilayer)
  WRITE(10,*) 'Material_Top(1,',ilayer,')=', Material_Top(1,ilayer)
 END DO

 ! For composite beam
 IF (Num_Material_Ele > 1) THEN
  DO COUNT_MAT = 2, Num_Material_Ele
   WRITE(*,*) 'Input the start and end coordinates of material',
   COUNT_MAT, ':'
   READ(*,*) START_CO(COUNT_MAT), END_CO(COUNT_MAT)
```

```
    DO ilayer = 1, Num_Layer_Material(COUNT_MAT)
     Material_Bot(COUNT_MAT,ilayer)=START_CO(COUNT_MAT)+(ilayer-1)*
(END_CO(COUNT_MAT)-START_CO(COUNT_MAT))/Num_Layer_Material
(COUNT_MAT)
     Material_Top(COUNT_MAT,ilayer) = START_CO(COUNT_MAT) + ilayer*
     (END_CO(COUNT_MAT)-START_CO(COUNT_MAT))/Num_Layer_Material
     (COUNT_MAT)
     WRITE(10,*) 'Material_Bot(',COUNT_MAT,ilayer,') =',
     Material_Bot(COUNT_MAT,ilayer)
     WRITE(10,*) 'Material_Top(',COUNT_MAT,ilayer,') =',
     Material_Top(COUNT_MAT,ilayer)
    END DO
   END DO
  END IF

  FE = 0
  FEy = 0
  DO i = 1, Num_Material_Ele
   DO j = 1, Num_Layer_Material(i)
    IF (i == 1) THEN
     FE = FE + Db(i)*BREADTH_LAYER(i,j)*(Material_Top(i,j)-
     Material_Bot(i,j))
     FEy = FEy + Db(i)*BREADTH_LAYER(i,j)*(Material_Top(i,j)-
Material_Bot(i,j))*(0.5*Thickness-(Material_Top(i,j)+Material_Bot
(i,j))/2)
    ELSE
     FE = FE + (Db(i)-Db(1))* BREADTH_LAYER(i,j)*(Material_Top(i,j)-
Material_Bot(i,j))
     FEy = FEy + (Db(i)-Db(1))* BREADTH_LAYER(i,j)*(Material_Top(i,j)-
Material_Bot(i,j))*(0.5*Thickness-(Material_Top(i,j)+Material_Bot
(i,j))/2)
    END IF
   END DO
  END DO

  ! Neutral axis
  NEU_AXIS = FEy/FE

  ! For homogeneous beam
  Dbb(1) = 0
  Dss(1) = 0
  Dxl_LIN(1) = 0
  Dxb_LIN(1) = 0

  DO ilayer = 1, Num_Layer_Material(1)
   Dbb(1) = Dbb(1) + BREADTH_LAYER(1,ilayer)*Db(1)*(Material_Top(1,
ilayer)**3-Material_Bot(1,ilayer)**3)/3
   Dss(1) = Dss(1) + BREADTH_LAYER(1,ilayer)*Ds(1)*(Material_Top(1,
ilayer)-Material_Bot(1,ilayer))*5/6
   Dxl_LIN(1) = Dxl_LIN(1) + BREADTH_LAYER(1,ilayer)*Db(1)*
(Material_Top(1,ilayer)-Material_Bot(1,ilayer))
```

```
  Dxb_LIN(1) = Dxb_LIN(1) + BREADTH_LAYER(1,ilayer)*Db(1)*
(Material_Top(1,ilayer)-Material_Bot(1,ilayer))*ABS((0.5*
Thickness-0.5*(Material_Top(1,ilayer)+Material_Bot(1,ilayer))-
NEU_AXIS))
  END DO

 ! For composite beam
 IF (Num_Material_Ele > 1) THEN
  DO COUNT_MAT = 2, Num_Material_Ele
   Dbb(COUNT_MAT) = 0
   Dss(COUNT_MAT) = 0
   Dxl_LIN(COUNT_MAT) = 0
   Dxb_LIN(COUNT_MAT) = 0

   DO ilayer = 1, Num_Layer_Material(COUNT_MAT)
    Dbb(COUNT_MAT)=Dbb(COUNT_MAT)+BREADTH_LAYER(COUNT_MAT,ilayer)*
Db(COUNT_MAT)*(Material_Top(COUNT_MAT,ilayer)**3-Material_Bot
(COUNT_MAT,ilayer)**3)/3
    Dss(COUNT_MAT)=Dss(COUNT_MAT)+BREADTH_LAYER(COUNT_MAT,ilayer)*
Ds(COUNT_MAT)*(Material_Top(COUNT_MAT,ilayer)-Material_Bot(COUN-
T_MAT,ilayer))*5/6
    Dxl_LIN(COUNT_MAT)=Dxl_LIN(COUNT_MAT)+BREADTH_LAYER(COUNT_MAT,
ilayer)*Db(COUNT_MAT)*(Material_Top(COUNT_MAT,ilayer) - Material_Bot
(COUNT_MAT,ilayer))
    Dxb_LIN(COUNT_MAT)=Dxb_LIN(COUNT_MAT)+BREADTH_LAYER(COUNT_MAT,
ilayer)*Db(COUNT_MAT)*(Material_Top(COUNT_MAT,ilayer) - Material_Bot
(COUNT_MAT,ilayer)) &
                * ABS(0.5*Thickness-NEU_AXIS-(Material_Top(COUNT_MAT,
ilayer)+Material_Bot(COUNT_MAT,ilayer))/2)
   END DO
  END DO
 END IF

 Bending_Stiff = 0
 Dxl_LIN_STIFF = 0
 Dxb_LIN_STIFF = 0

 DO COUNT_MAT = 1, Num_Material_Ele
  Bending_Stiff = Bending_Stiff + Dbb(COUNT_MAT)
  Dxl_LIN_STIFF = Dxl_LIN_STIFF + Dxl_LIN(COUNT_MAT)
  Dxb_LIN_STIFF = Dxb_LIN_STIFF + Dxb_LIN(COUNT_MAT)
  Shear_Stiff = Shear_Stiff + Dss(COUNT_MAT)
 END DO

END SUBROUTINE ELE_STIFF_LINEAR

SUBROUTINE Shape_Function_Linear
!------------------------------------------------------------------------------
! The subroutine Shape_Function_Linear provides
! shape function for beam element
!------------------------------------------------------------------------------
```

```
  USE MODULE_NAME
  IMPLICIT NONE

  Qb = Bending_Stiff
  Qs = Shear_Stiff

  Miu = 1/(1+12*Qb/(Qs*L**2))
  Shape_Bb(1,1) = 6*Miu*(1/L-2*x/L**2)/L
  Shape_Bb(1,2) = (3*Miu+1)/L-6*Miu*x/L**2
  Shape_Bb(1,3) = -6*Miu*(1/L-2*x/L**2)/L
  Shape_Bb(1,4) = -(1-3*Miu)/L-6*Miu*x/L**2
  Shape_Bs(1,1) = -1/L+Miu/L
  Shape_Bs(1,2) = Miu/2-1.0/2
  Shape_Bs(1,3) = 1/L-Miu/L
  Shape_Bs(1,4) = -1.0/2+Miu/2

 END SUBROUTINE Shape_Function_Linear


 SUBROUTINE Gauss_Beam(POSGP,WEIGP)
 !-----------------------------------------------------------------------------------
 ! The subroutine Gauss_Beam gives the gauss
 ! integral constants for beam element
 !-----------------------------------------------------------------------------------
  USE MODULE_NAME
  IMPLICIT NONE

  ! Declare local variables
  REAL, ALLOCATABLE, DIMENSION(:) :: POSGP, WEIGP

  WRITE(*,*) 'Input the ordinates of Gaussian integration'
  READ(*,*) Num_Gauss_Point

  ! Allocate array dimension for Gaussian points
  ALLOCATE(POSGP(Num_Gauss_Point),WEIGP(Num_Gauss_Point))

  POSGP = 0.0
  WEIGP = 0.0

  SELECT CASE(Num_Gauss_Point)
   CASE(1)
    POSGP(1)=0.0
    WEIGP(1)=2.0
   CASE(2)
    POSGP(1)=-0.577350269189626
    POSGP(2)=-POSGP(1)
    WEIGP(1)=1.0
    WEIGP(2)=1.0
   CASE(3)
    POSGP(1)=-0.774596669241483
    POSGP(2)=0.0
```

```
   POSGP(3)=-POSGP(1)
   WEIGP(1)=0.555555555555556
   WEIGP(2)=0.888888888888889
   WEIGP(3)=WEIGP(1)
  CASE(4)
   POSGP(1)=-0.861136311594053
   POSGP(2)=-0.339981043584856
   POSGP(3)=-POSGP(2)
   POSGP(4)=-POSGP(1)
   WEIGP(1)=0.347854845137454
   WEIGP(2)=0.652145154862546
   WEIGP(3)=WEIGP(2)
   WEIGP(4)=WEIGP(1)
  CASE(5)
   POSGP(1)=-0.906179845938664
   POSGP(2)=-0.538469310105683
   POSGP(3)=0.0
   POSGP(4)=-POSGP(2)
   POSGP(5)=-POSGP(1)
   WEIGP(1)=0.236926885056189
   WEIGP(2)=0.478628670499366
   WEIGP(3)=0.568888888888889
   WEIGP(4)=WEIGP(2)
   WEIGP(5)=WEIGP(1)
  CASE(6)
   POSGP(1)=-0.932469514203152
   POSGP(2)=-0.661209386466265
   POSGP(3)=-0.238619186083197
   POSGP(4)=-POSGP(3)
   POSGP(5)=-POSGP(2)
   POSGP(6)=-POSGP(1)
   WEIGP(1)=0.171324492379170
   WEIGP(2)=0.360761573048139
   WEIGP(3)=0.467913934572691
   WEIGP(4)=WEIGP(3)
   WEIGP(5)=WEIGP(2)
   WEIGP(6)=WEIGP(1)
  CASE DEFAULT
   WRITE(*,*)' Sorry, you have type a wrong gauss point number!'
   WRITE(4,*)' Sorry, you have type a wrong gauss point number!'
  STOP
  END SELECT

END SUBROUTINE Gauss_Beam


SUBROUTINE GAUSS_ELIMINATION
!-----------------------------------------------------------------------------
! The subroutine GAUSS_ELIMINATION finds Gauss
! elimination
```

```
!-----------------------------------------------------------------------------
 USE MODULE_NAME
 IMPLICIT NONE
 INTEGER :: COUNT1, COUNT2
 REAL :: R, XMULT

 SOLVBL = .TRUE.
 ROW = 1

 DO WHILE (SOLVBL .AND. ROW < Num_Unknown)
  CALL PIVOT(SOLVBL)

  IF (SOLVBL) THEN
   R = 1.0/AUG(ROW,ROW)
   AUG(ROW,ROW) = 1.0

   DO COUNT1 = ROW+1,Num_Unknown+1
    AUG(ROW,COUNT1) = AUG(ROW,COUNT1)*R
   END DO

   DO COUNT2 = ROW+1,Num_Unknown
    XMULT = -AUG(COUNT2,ROW)
    AUG(COUNT2,ROW) = 0.0

    DO COUNT1 = ROW+1,Num_Unknown+1
     AUG(COUNT2,COUNT1) = AUG(COUNT2,COUNT1) + XMULT*AUG(ROW,COUNT1)
    END DO
   END DO
  END IF

  ROW = ROW+1
 END DO

 IF (SOLVBL) THEN
  R = 1.0/AUG(Num_Unknown,Num_Unknown)
  AUG(Num_Unknown,Num_Unknown) = 1.0
  AUG(Num_Unknown,Num_Unknown+1)=AUG(Num_Unknown,Num_Unknown+1)*R
 END IF

END SUBROUTINE GAUSS_ELIMINATION


SUBROUTINE PIVOT(FOUND)
!-----------------------------------------------------------------------------
! The subroutine PIVOT finds PIVOT for Gauss
! elimination
!-----------------------------------------------------------------------------
 USE MODULE_NAME
 IMPLICIT NONE
 REAL :: XMAX, XTEMP
```

```
 LOGICAL :: FOUND
 INTEGER :: MAXROW, COUNT1, COUNT2

 XMAX = ABS(AUG(ROW,ROW))
 MAXROW = ROW

 DO COUNT1 = ROW+1, Num_Unknown
  IF (ABS(AUG(COUNT1,ROW)) > XMAX) THEN
   XMAX = ABS(AUG(COUNT1,ROW))
   MAXROW = COUNT1
  END IF
 END DO

 IF (XMAX == 0) THEN
  FOUND = .FALSE.
 ELSE IF (MAXROW /= ROW) THEN
  DO COUNT2 = ROW, Num_Unknown+1
   XTEMP = AUG(ROW,COUNT2)
   AUG(ROW,COUNT2) = AUG(MAXROW,COUNT2)
   AUG(MAXROW,COUNT2) = XTEMP
  END DO
 END IF

END SUBROUTINE PIVOT


SUBROUTINE BACK(Unknown)
!------------------------------------------------------------
! The subroutine BACK returns solution for Gauss
! Elimination
!------------------------------------------------------------
 USE MODULE_NAME
 IMPLICIT NONE
 INTEGER :: COUNT1, COUNT2
 REAL :: SUM
 REAL, ALLOCATABLE, DIMENSION(:,:) :: Unknown

 Unknown(Num_Unknown,1) = AUG(Num_Unknown,Num_Unknown+1)

 DO COUNT1 = Num_Unknown-1, 1, -1
  SUM = 0.0

  DO COUNT2 = COUNT1+1, Num_Unknown
   SUM = SUM + AUG(COUNT1,COUNT2)*Unknown(COUNT2,1)
  END DO

  Unknown(COUNT1,1) = AUG(COUNT1,Num_Unknown+1) - SUM
 END DO

END SUBROUTINE BACK
```

Appendix G: Finite element code for composite beam element: Nonlinear analysis

This appendix provides the principal finite element code for nonlinear analysis of reinforced concrete beams using the 1-D composite beam element introduced in this book. The main program and the subroutines for linear analysis can be referred to Appendix F. The finite element code is written in Fortran 90.

```
PROGRAM REINFORCED_CONCRETE_BEAM
!---------------------------------------------------------------------------------
! This program is used for nonlinear analysis of
! reinforced concrete beams.
!---------------------------------------------------------------------------------

{The main program is the same as that for the linear analysis in
Appendix F. Omitted herein}

END PROGRAM REINFORCED_CONCRETE_BEAM


SUBROUTINE ITERATION_NEWTON
!---------------------------------------------------------------------------------
! The subroutine ITERATION_NEWTON finds
! nonlinear solution.
!---------------------------------------------------------------------------------
 USE MODULE_NAME
 IMPLICIT NONE
 INTEGER :: i, j, m, RANGE, COUNT_Z
 REAL, ALLOCATABLE, DIMENSION(:,:) :: Incremental_Total

 ALLOCATE(Incremental_Total(Num_Unknown,1))

 COUNT_ITER = 0
 DO
  CALL NEWTON_RAPHSON

  EXIT_CONDITION = 0
  DO COUNT_NR = 1, Num_Unknown
   Incremental_Total(COUNT_NR,1) = Displacement_Vec(COUNT_NR,1) -
Displacement_Initial(COUNT_NR,1)
```

```
   COMPARE=ABS(Displacement_Increment(COUNT_NR,1)/Incremental_Total
(COUNT_NR,1))
   IF (COMPARE > TOL) THEN
    EXIT_CONDITION = EXIT_CONDITION+1
   END IF
  END DO

  COUNT_ITER = COUNT_ITER + 1

  DEALLOCATE(Displacement_Increment)
  DEALLOCATE(STRAIN_X,STRAIN_Z)
  DEALLOCATE(LOCATE_Z)
  DEALLOCATE(Dbb_MAR,Dss_MAR)
  DEALLOCATE(K_Large_Global_Bound,K_Sigma_Global_Bound)
  DEALLOCATE(K_Large_Global,K_Sigma_Global)
  DEALLOCATE(Dxl_MAR,Dxb_MAR)

  IF (EXIT_CONDITION == 0) THEN
   WRITE(*,*) 'SATISFY THE CONVERGENCE CONDITION'
   EXIT
  ELSE IF (COUNT_ITER == MAXITER) THEN
   WRITE(*,*) 'REACH THE MAXIMUM ITERATION NUMBER'
   EXIT
  END IF
 END DO

 DEALLOCATE(Incremental_Total)

END SUBROUTINE ITERATION_NEWTON


SUBROUTINE NEWTON_RAPHSON
!-------------------------------------------------------------------------------
! The subroutine NEWTON_RAPHSON finds
! displacement increment.
!-------------------------------------------------------------------------------
 USE MODULE_NAME
 IMPLICIT NONE
 INTEGER :: COUNT_i, COUNT_j, COUNT_NUM, CC, i, COUNT_LAYER
 INTEGER, ALLOCATABLE, DIMENSION(:) :: INDX

 CALL MARNON_STRAIN
 CALL EFF_INERTIA
 CALL INTERNAL_FORCE
 CALL ELE_STIFF_MARNON
 CALL Global_Stiff_Matrix_MARNON
 CALL Global_Stiff_Matrix_GeoNon
 CALL Global_Boundary

 ALLOCATE(NR_COE(Num_Unknown ,1))
 ALLOCATE(NR_K(Num_Unknown,Num_Unknown), NR_K1(DIMENSION_GLOBAL,
Num_Unknown))
```

```
NR_K = 0
NR_K1 = 0
DO COUNT_i = 1, DIMENSION_GLOBAL
 SKIP = 0
 DO COUNT_j = 1, DIMENSION_GLOBAL
  NONZERO = 0
  DO COUNT_NUM = 1, C_0
   IF (COUNT_j == COLUMN_ZERO(COUNT_NUM)) THEN
    NONZERO = NONZERO + 1
   END IF
  END DO
  IF (NONZERO == 0) THEN
   SKIP = SKIP+1
   NR_K1(COUNT_i,SKIP) = K_Global_Bound(COUNT_i,COUNT_j)
  END IF
 END DO
END DO

DO COUNT_j = 1, Num_Unknown
 SKIP = 0
 DO COUNT_i = 1, DIMENSION_GLOBAL
  NONZERO = 0
  DO COUNT_NUM = 1, R_0
   IF (COUNT_i == ROW_ZERO(COUNT_NUM)) THEN
    NONZERO = NONZERO + 1
   END IF
  END DO
  IF (NONZERO == 0) THEN
   SKIP = SKIP+1
   NR_K(SKIP,COUNT_j) = NR_K1(COUNT_i,COUNT_j)
  END IF
 END DO
END DO

ALLOCATE(NR_LOAD(Num_Unknown,1))

SKIP = 0
DO COUNT_i = 1,DIMENSION_GLOBAL
 NONZERO = 0
 DO COUNT_NUM = 1, R_0
  IF (COUNT_i == ROW_ZERO(COUNT_NUM)) THEN
   NONZERO = NONZERO + 1
  END IF
 END DO
 IF (NONZERO == 0) THEN
  SKIP = SKIP+1
  NR_LOAD(SKIP,1) = Nodal_Load_Non(COUNT_i,1)
 END IF
END DO
```

```
  NR_COE = NR_LOAD - INTER_FORCE

  ALLOCATE(Displacement_Increment(Num_Unknown,1))
  ALLOCATE(NR_invK(Num_Unknown,Num_Unknown))
  ALLOCATE(INDX(Num_Unknown))

  CALL MIGS(NR_K, Num_Unknown, NR_invK, INDX)

  Displacement_Increment = MATMUL(NR_invK,NR_COE)
  Displacement_Vec = Displacement_Vec + Displacement_Increment

  CC = 0
  DO COUNT_i = 1, DIMENSION_GLOBAL
   IF (SUPPORT(COUNT_i,1) == 0) THEN
    Displacement_Global(COUNT_i,1) = 0
    Increment_Global(COUNT_i,1) = 0
   ELSE
    CC = CC + 1
    Displacement_Global(COUNT_i,1) = Displacement_Vec(CC,1)
    Increment_Global(COUNT_i,1) = Displacement_Increment(CC,1)
   END IF
  END DO

  DO COUNT_j = 1, Num_Node_Stru
   DIS_Global(COUNT_j,1) = Displacement_Global(2*COUNT_j-1,1)
   ROTA_Global(COUNT_j,1) = Displacement_Global(2*COUNT_j,1)
  END DO

  DEALLOCATE(NR_COE,NR_K,NR_K1,NR_invK,INDX)
  DEALLOCATE(NR_LOAD)
  DEALLOCATE(FEy_CRACK,FE_CRACK)

 END SUBROUTINE NEWTON_RAPHSON


 SUBROUTINE MARNON_STRAIN
 !------------------------------------------------------------------------------
 ! The subroutine MARNON_STRAIN finds strains
 ! at Gauss Point.
 !------------------------------------------------------------------------------
  USE MODULE_NAME
  IMPLICIT NONE
  REAL, ALLOCATABLE, DIMENSION(:) :: u, w
  INTEGER :: NUM_GAUSS_STRAINPOINT
  INTEGER :: COUNT_Gauss, COUNT_DIMN, COUNT_LAYER, COUNT
  INTEGER :: i, j, m, n, COUNT_SHAPE, COUNT_SS, COUNT_ELE
  REAL :: ELE_X

  ALLOCATE(STRAIN_X(Num_Material_Ele,DIMN_STRAIN),STRAIN_Z
 (Num_Material_Ele,DIMN_STRAIN))
  ALLOCATE(LOCATE_Z(Num_Material_Ele,DIMN_STRAIN))
```

```
! Call subroutine to select gauss integral constants
CALL Gauss_Beam(u,w)

DO i = 1, Num_Material_Ele
 DO j = 1, Num_Element_Stru
  DO COUNT_Gauss = 1, Num_Gauss_Point
   DO COUNT_LAYER = 1, Num_Layer_Material(i)
    COUNT_DIMN = Num_Gauss_Point*Num_Layer_Material(i)*(j-1) +
Num_Layer_Material(i)*(COUNT_Gauss-1) + COUNT_LAYER
    ELE_X = 0.5*L*u(COUNT_Gauss) + 0.5*L
    STRAIN_X(i,COUNT_DIMN) = ELE_X + (j-1)*L
    STRAIN_Z(i,COUNT_DIMN) = 0.5 * (Material_Bot(i,COUNT_LAYER)+
Material_Top(i,COUNT_LAYER))
   END DO
  END DO
 END DO
END DO

COUNT = 0
INCREMENT_Bl_STRAIN = 0
INCREMENT_Bb_STRAIN = 0
DO i = 1, Num_Material_Ele
 COUNT_SHAPE = 0
 COUNT_SS = 0
 COUNT_ELE = 0
 DO j = 1, Num_Gauss_Point*Num_Layer_Material(i)*Num_Element_Stru
  COUNT = COUNT + 1
  COUNT_SS = COUNT_SS + 1
  COUNT_SHAPE = (COUNT_SS-1)/Num_Layer_Material(i) + 1
  COUNT_ELE=(COUNT_SS-1)/(Num_Layer_Material(i)*Num_Gauss_Point)+1

  m = (INT(STRAIN_X(i,j)/L) + 1) * 2 - 1
  DO n = 1, Dimension_Ele
   Displacement_Vec_Ele(n,1) = Displacement_Global(m,1)
   Increment_Vec_Ele(n,1) = Increment_Global(m,1)
   m = m + 1
  END DO

  x = STRAIN_X(i,j) - INT(STRAIN_X(i,j)/L) * L

  IF (STEP_NON == 1) THEN
   IF (COUNT_ITER == 0) THEN
    CALL Shape_Function_Linear
    CALL Shape_Function_Nonlinear(COUNT_SHAPE)
    DO n = 1, Dimension_Ele
     Bending_Strain_CHECK(i,j) = Bending_Strain_CHECK(i,j) +
Shape_Bb(1,n) * Displacement_Vec_Ele(n,1)
     Shear_Strain_CHECK(i,j) = Shear_Strain_CHECK(i,j) +
Shape_Bs(1,n)* Displacement_Vec_Ele(n,1)
     Large_Strain(i,j) = 0.5 * Shape_Bl(COUNT_ELE,n) *
Displacement_Vec_Ele(n,1)
    END DO
```

```
     IF (NEU_AXIS == 0.5*Thickness) THEN
      IF (Bending_Strain_CHECK(i,j) > 0) THEN
       STRAIN_Z(i,j) = -STRAIN_Z(i,j)
      END IF

      DO n = 1, Dimension_Ele
       Bending_Strain_CHECK_Z(i,j) = Bending_Strain_CHECK_Z(i,j) +
(STRAIN_Z(i,j) * Shape_Bb(1,n) + 0.5 * Shape_Bl(COUNT_ELE,n)) *
Displacement_Vec_Ele(n,1)
      END DO
     END IF

     IF (NEU_AXIS /= 0.5*Thickness) THEN
      LOCATE_Z(i,j) = 0.5*Thickness-STRAIN_Z(i,j)-NEU_AXIS

      IF (Bending_Strain_CHECK(i,j) < 0) THEN
        LOCATE_Z(i,j) = - LOCATE_Z(i,j)
      END IF

      DO n = 1, Dimension_Ele
       Bending_Strain_CHECK_Z(i,j) = Bending_Strain_CHECK_Z(i,j) +
(LOCATE_Z(i,j) * Shape_Bb(1,n) + 0.5 * Shape_Bl(COUNT_ELE,n)) *
Displacement_Vec_Ele(n,1)
      END DO

     END IF

    ELSE
     CALL Shape_Function_MARNON(COUNT_SHAPE)
     CALL Shape_Function_Nonlinear(COUNT_SHAPE)
     DO n = 1, Dimension_Ele
      Bending_Strain_CHECK(i,j) = Bending_Strain_CHECK(i,j) +
Shape_Bb(1,n) * Increment_Vec_Ele(n,1)
      Shear_Strain_CHECK(i,j) = Shear_Strain_CHECK(i,j) +
Shape_Bs(1,n) * Increment_Vec_Ele(n,1)
      INCREMENT_Bl_STRAIN(i,j) = INCREMENT_Bl_STRAIN(i,j) +
Shape_Bl(COUNT_ELE,n) * Increment_Vec_Ele(n,1)
      INCREMENT_Bb_STRAIN(i,j) = INCREMENT_Bb_STRAIN(i,j) +
Shape_Bb(1,n) * Increment_Vec_Ele(n,1)
     END DO

     IF (AXIS_CRACK(COUNT_SHAPE) == 0.5*Thickness) THEN
      IF (Bending_Strain_CHECK(i,j) > 0) THEN
       STRAIN_Z(i,j) = -STRAIN_Z(i,j)
      END IF

      DO n = 1, Dimension_Ele
       Bending_Strain_CHECK_Z(i,j) = Bending_Strain_CHECK_Z(i,j) +
(STRAIN_Z(i,j) * Shape_Bb(1,n) + Shape_Bl(COUNT_ELE,n)) *
Increment_Vec_Ele(n,1)
      END DO
     END IF
```

```
     IF (AXIS_CRACK(COUNT_SHAPE) /= 0.5*Thickness) THEN
      LOCATE_Z(i,j) = 0.5*Thickness-STRAIN_Z(i,j)-AXIS_CRACK
(COUNT_SHAPE)

      IF (Bending_Strain_CHECK(i,j) < 0) THEN
       LOCATE_Z(i,j) = - LOCATE_Z(i,j)
      END IF

      DO n = 1, Dimension_Ele
       Bending_Strain_CHECK_Z(i,j) = Bending_Strain_CHECK_Z(i,j) +
(LOCATE_Z(i,j) * Shape_Bb(1,n) + Shape_Bl(COUNT_ELE,n)) *
Increment_Vec_Ele(n,1)
      END DO
     END IF

    END IF
   END IF

   IF (STEP_NON > 1) THEN
    CALL Shape_Function_MARNON(COUNT_SHAPE)
    CALL Shape_Function_Nonlinear(COUNT_SHAPE)
    DO n = 1, Dimension_Ele
     Bending_Strain_CHECK(i,j) = Bending_Strain_CHECK(i,j) +
Shape_Bb(1,n) * Increment_Vec_Ele(n,1)
     Shear_Strain_CHECK(i,j) = Shear_Strain_CHECK(i,j) +
Shape_Bs(1,n) * Increment_Vec_Ele(n,1)
     INCREMENT_Bl_STRAIN(i,j) = INCREMENT_Bl_STRAIN(i,j) + Shape_Bl
(COUNT_ELE,n) * Increment_Vec_Ele(n,1)
     INCREMENT_Bb_STRAIN(i,j) = INCREMENT_Bb_STRAIN(i,j) +
Shape_Bb(1,n) * Increment_Vec_Ele(n,1)
    END DO

    IF (AXIS_CRACK(COUNT_SHAPE) == 0.5*Thickness) THEN
     IF (Bending_Strain_CHECK(i,j) > 0) THEN
      STRAIN_Z(i,j) = -STRAIN_Z(i,j)
     END IF

     DO n = 1, Dimension_Ele
      Bending_Strain_CHECK_Z(i,j) = Bending_Strain_CHECK_Z(i,j) +
(STRAIN_Z(i,j) * Shape_Bb(1,n) + Shape_Bl(COUNT_ELE,n)) *
Increment_Vec_Ele(n,1)
     END DO
    END IF

    IF (AXIS_CRACK(COUNT_SHAPE) /= 0.5*Thickness) THEN
     LOCATE_Z(i,j) = 0.5*Thickness-STRAIN_Z(i,j)-AXIS_CRACK
(COUNT_SHAPE)

     IF (Bending_Strain_CHECK(i,j) < 0) THEN
      LOCATE_Z(i,j) = - LOCATE_Z(i,j)
     END IF
```

```
    DO n = 1, Dimension_Ele
     Bending_Strain_CHECK_Z(i,j) = Bending_Strain_CHECK_Z(i,j) +
(LOCATE_Z(i,j) * Shape_Bb(1,n) + Shape_Bl(COUNT_ELE,n)) *
Increment_Vec_Ele(n,1)
    END DO
   END IF
  END IF

  Strain_PRINCIPAL(i,j) = Bending_Strain_PRINCIPAL(i,j)
  IF (Bending_Strain_CHECK_Z(i,j) >= 0) THEN
  Bending_Strain_PRINCIPAL(i,j)=Bending_Strain_CHECK_Z(i,j)/2+SQRT
((Bending_Strain_CHECK_Z(i,j)/2)**2+(Shear_Strain_CHECK(i,j)/2)**2)
  ELSE
   Bending_Strain_PRINCIPAL(i,j) = Bending_Strain_CHECK_Z(i,j)/2 -
SQRT((Bending_Strain_CHECK_Z(i,j)/2)**2+(Shear_Strain_CHECK(i,
j)/2)**2)
  END IF

  DEALLOCATE(Shape_Bl)
 END DO
END DO

END SUBROUTINE MARNON_STRAIN


SUBROUTINE ELE_STIFF_MARNON
!-------------------------------------------------------------------------------
! The subroutine ELE_STIFF_MARNON generates
! element stiffness for nonlinear analysis.
!-------------------------------------------------------------------------------
 USE MODULE_NAME
 IMPLICIT NONE
 INTEGER :: COUNT, COUNT_MAR, COUNT_LAYER, COUNT_GAUSS, COUNT_R,
COUNT_C, MAX, i, j, k
 REAL :: STRAIN_1, STRAIN_2, EFF_HEIGHT, OPP_HEIGHT

 ALLOCATE(Dbb_MAR(Num_Material_Ele,Num_Gauss_Point*
Num_Element_Stru))
 ALLOCATE(Dss_MAR(Num_Material_Ele,Num_Gauss_Point*
Num_Element_Stru))
 ALLOCATE(Dxl_MAR(Num_Material_Ele,Num_Gauss_Point*
Num_Element_Stru))
 ALLOCATE(Dxb_MAR(Num_Material_Ele,Num_Gauss_Point*
Num_Element_Stru))

 IF (STEP_NON ==1) THEN
  IF (COUNT_ITER == 0) THEN
```

```
    DO k = 1, Num_Gauss_Point*Num_Element_Stru
     AXIS_CRACK(k) = NEU_AXIS
    END DO

    REO_RATIO = AREA_TOTAL/Aeff
    EFF_HEIGHT = Aeff/breadth
    OPP_HEIGHT = Thickness - EFF_HEIGHT

   END IF
  END IF

  ! Concrete is the 1st material type by default
  COUNT_LAYER = 0
  COUNT_GAUSS = 0
  DO j =1, Num_Gauss_Point*Num_Layer_Material(1)*Num_Element_Stru
   COUNT_GAUSS = INT((j-1)/Num_Layer_Material(1)) + 1
   COUNT_LAYER = j - (COUNT_GAUSS-1)*Num_Layer_Material(1)

   IF (Bending_Strain_PRINCIPAL(1,j) >= 0) THEN
    IF (Bending_Strain_PRINCIPAL(1,j) > Crack_Strain(1)) THEN
     STIFFENING_FACTOR  =  EXP(-1100*(Bending_Strain_PRINCIPAL(1,j)-
Crack_Strain(1))*(EA_TOTAL/(200*AREA_TOTAL)))
     STRAIN_1 = (1.3-STIFFENING_FACTOR)*Crack_Stress(1)*
(1/(REO_RATIO*EA_TOTAL/AREA_TOTAL)+1/Db(1))
     STRAIN_2 = STRAIN_BAR - STIFFENING_FACTOR*Crack_Stress(1)*
(1/(REO_RATIO*EA_TOTAL/AREA_TOTAL)+1/Db(1))

     IF (Bending_Strain_PRINCIPAL(1,j) < STRAIN_1) THEN
      Db_MAR(1,j) = Crack_Stress(1)*(1-(1-STIFFENING_FACTOR)*
Bending_Strain_PRINCIPAL(1,j)/STRAIN_1)/
Bending_Strain_PRINCIPAL(1,j)
     ELSE IF (Bending_Strain_PRINCIPAL(1,j) < STRAIN_2) THEN
      Db_MAR(1,j) = STIFFENING_FACTOR*Crack_Stress(1)/
Bending_Strain_PRINCIPAL(1,j)
     ELSE IF (Bending_Strain_PRINCIPAL(1,j) < STRAIN_BAR) THEN
      Db_MAR(1,j)  =  STIFFENING_FACTOR*Crack_Stress(1)*((STRAIN_BAR-
Bending_Strain_PRINCIPAL(1,j))/(STRAIN_BAR-STRAIN_2))/
Bending_Strain_PRINCIPAL(1,j)
     ELSE
      Db_MAR(1,j) = 0
     END IF

     IF ((0.5*Thickness-(Material_Top(1,COUNT_LAYER) + Material_Bot
(1,COUNT_LAYER))/2) < OPP_HEIGHT) THEN
      Db_MAR(1,j) = 0
     END IF

    END IF
   END IF
```

```
 IF (Bending_Strain_PRINCIPAL(1,j) < 0) THEN
  IF (ABS(Bending_Strain_PRINCIPAL(1,j)) <= Compression_Strain
(1)) THEN
   Db_MAR(1,j)=0.85*Ulti_Stress(1)*(2-ABS(Bending_Strain_PRINCIPAL
(1,j))/Compression_Strain(1))/Compression_Strain(1)
  ELSE IF (ABS(Bending_Strain_PRINCIPAL(1,j)) <= 0.003) THEN ! Set
failure strain of concrete as 0.003
   Db_MAR(1,j) = 0.85*Ulti_Stress(1)*(1-0.15*((ABS
(Bending_Strain_PRINCIPAL(1,j))-Compression_Strain(1))/(0.003-
Compression_Strain(1))))/ABS(Bending_Strain_PRINCIPAL(1,j))
  ELSE
   Db_MAR(1,j) = 0
  END IF
 END IF

 IF (Bending_Strain_PRINCIPAL(1,j) >= 0) THEN
  IF (Bending_Strain_PRINCIPAL(1,j) > Crack_Strain(1)) THEN
   IF (Bending_Strain_PRINCIPAL(1,j) < 0.004) THEN
    Ds_MAR(1,j)=0.25*Ds(1)*(1-Bending_Strain_PRINCIPAL(1,j)/0.004)
   ELSE
    Ds_MAR(1,j) = 0
   END IF
  END IF
 END IF
END DO

COUNT = 0
DO i = 1, Num_Gauss_Point*Num_Element_Stru
 Dbb_MAR(1,i) = 0
 Dss_MAR(1,i) = 0
 Dxl_MAR(1,i) = 0
 Dxb_MAR(1,i) = 0

 DO j = 1, Num_Layer_Material(1)
  COUNT = COUNT + 1
  Dbb_MAR(1,i) = Dbb_MAR(1,i) + breadth*Db_MAR(1,COUNT)*
(Material_Top(1,j)-Material_Bot(1,j))*(0.5*Thickness-AXIS_CRACK
(i)-0.5*(Material_Top(1,j)+Material_Bot(1,j)))**2
  Dss_MAR(1,i) = Dss_MAR(1,i) + breadth*Ds_MAR(1,COUNT)*
(Material_Top(1,j)-Material_Bot(1,j))*5/6
  Dxl_MAR(1,i) = Dxl_MAR(1,i) + breadth*Db_MAR(1,COUNT)*
(Material_Top(1,j)-Material_Bot(1,j))
  Dxb_MAR(1,i) = Dxb_MAR(1,i) + breadth*Db_MAR(1,COUNT)*
(Material_Top(1,j)-Material_Bot(1,j))*ABS((0.5*Thickness-0.5*
(Material_Top(1,j)+Material_Bot(1,j))-AXIS_CRACK(i)))
 END DO

END DO

! Reinforcing materials start from here
DO i = 2, Num_Material_Ele
```

```
  DO j =1, Num_Gauss_Point*Num_Layer_Material(i)*Num_Element_Stru
   IF (Crack_Stress(i) /= 0) THEN ! FRP
    Crack_Strain(i) = Crack_Stress(i)/Db(i)

    IF (Bending_Strain_PRINCIPAL(i,j) >= 0) THEN
     IF (Bending_Strain_PRINCIPAL(i,j) > Crack_Strain(i)) THEN
      Db_MAR(i,j) = 0
     ENDIF
    ENDIF

    IF (Bending_Strain_PRINCIPAL(i,j) < 0) THEN
     Db_MAR(i,j) = 0
    END IF
   END IF

   IF (Crack_Stress(i) == 0) THEN ! STEEL
    Compression_Strain(i) = Ulti_Stress(i)/Db(i)

    IF (Bending_Strain_PRINCIPAL(i,j) >= 0) THEN
     IF (Bending_Strain_PRINCIPAL(i,j) > Tension_Strain(i)) THEN
      Db_MAR(i,j) = Ulti_Stress(i)/Bending_Strain_PRINCIPAL(i,j)
     END IF
    END IF

    IF (Bending_Strain_PRINCIPAL(i,j) < 0) THEN
     IF(ABS(Bending_Strain_PRINCIPAL(i,j)) > Compression_Strain(i))
THEN
      Db_MAR(i,j)=Ulti_Stress(i)/ABS(Bending_Strain_PRINCIPAL(i,j))
     END IF
    END IF
   END IF

  END DO
 END DO

 Db_MAR_CRACK = Db_MAR

 DO COUNT_MAR = 2, Num_Material_Ele
  COUNT_R = 0
  DO k = 1, Num_Gauss_Point*Num_Element_Stru
   DO i = 1, Num_Layer_Material(COUNT_MAR)
    COUNT_R = COUNT_R + 1
    COUNT_C = (k-1)*Num_Layer_Material(1)
    DO j = 1, Num_Layer_Material(1)
     COUNT_C = COUNT_C + 1
     IF ((Material_Top(COUNT_MAR,i) + Material_Bot(COUNT_MAR,i))/
2 >= Material_Bot(1,j)) THEN
      IF ((Material_Top(COUNT_MAR,i) + Material_Bot(COUNT_MAR,i))/
2 <= Material_Top(1,j)) THEN
       Db_MAR_CRACK(COUNT_MAR,COUNT_R) = Db_MAR(COUNT_MAR,COUNT_R) -
Db_MAR(1,COUNT_C)
      END IF
```

```
    END IF
   END DO
  END DO
 END DO
END DO

DO COUNT_MAR = 2, Num_Material_Ele
 COUNT = 0
 DO i = 1, Num_Gauss_Point*Num_Element_Stru
  Dbb_MAR(COUNT_MAR,i) = 0
  Dss_MAR(COUNT_MAR,i) = 0
  Dxl_MAR(COUNT_MAR,i) = 0
  Dxb_MAR(COUNT_MAR,i) = 0
  DO j = 1, Num_Layer_Material(COUNT_MAR)
   COUNT = COUNT + 1
   Dbb_MAR(COUNT_MAR,i)=Dbb_MAR(COUNT_MAR,i)+breadth*Db_MAR_CRACK
(COUNT_MAR,COUNT)*(Material_Top(COUNT_MAR,j) - Material_Bot
(COUNT_MAR,j))*(0.5*Thickness-AXIS_CRACK(i)-(Material_Top
(COUNT_MAR,j)+Material_Bot(COUNT_MAR,j))/2)**2
   Dxl_MAR(COUNT_MAR,i)=Dxl_MAR(COUNT_MAR,i)+breadth*Db_MAR_CRACK
(COUNT_MAR,COUNT)*(Material_Top(COUNT_MAR,j) - Material_Bot
(COUNT_MAR,j))
   Dxb_MAR(COUNT_MAR,i)=Dxb_MAR(COUNT_MAR,i)+breadth*Db_MAR_CRACK
(COUNT_MAR,COUNT)*(Material_Top(COUNT_MAR,j) - Material_Bot
(COUNT_MAR,j))* ABS(0.5*Thickness-AXIS_CRACK(i)-(Material_Top
(COUNT_MAR,j)+Material_Bot(COUNT_MAR,j))/2)
  END DO
 END DO
END DO

Bending_Stiff_MAR = 0
Dxl_STIFF = 0
Dxb_STIFF = 0
DO i = 1, Num_Gauss_Point*Num_Element_Stru
 DO COUNT_MAR = 1, Num_Material_Ele
  Bending_Stiff_MAR(i)=Bending_Stiff_MAR(i)+Dbb_MAR(COUNT_MAR,i)
  Dxl_STIFF(i) = Dxl_STIFF(i) + Dxl_MAR(COUNT_MAR,i)
  Dxb_STIFF(i) = Dxb_STIFF(i) + Dxb_MAR(COUNT_MAR,i)
 END DO
  Shear_Stiff_MAR(i) = Dss_MAR(1,i)
END DO

END SUBROUTINE ELE_STIFF_MARNON


SUBROUTINE Global_Stiff_Matrix_MARNON
!---------------------------------------------------------------------------
! The subroutine Global_Stiff_Matrix_MARNON
! generates global stiffness matrix.
!---------------------------------------------------------------------------
```

```
 USE MODULE_NAME
 IMPLICIT NONE
 INTEGER :: COUNT_DIMEN, COUNT_i, COUNT_j, locate

 locate = 0
 Kb_Global = 0
 Ks_Global = 0
 DO COUNT_DIMEN = 1, Num_Element_Stru
  CALL STIFF_MATRIX_MARNON(COUNT_DIMEN)
  DO COUNT_i = 1, 4
   DO COUNT_j = 1, 4
    Kb_Global(COUNT_i+locate,COUNT_j+locate) = Kb_Global(COUNT_i+
locate,COUNT_j+locate) + Kb(COUNT_i,COUNT_j)
    Ks_Global(COUNT_i+locate,COUNT_j+locate) = Ks_Global(COUNT_i+
locate,COUNT_j+locate) + Ks(COUNT_i,COUNT_j)
   END DO
  END DO
  locate = locate + 2
 END DO

 DO COUNT_i = 1, DIMENSION_GLOBAL
  DO COUNT_j = 1, DIMENSION_GLOBAL
   K_Global(COUNT_i,COUNT_j) = Kb_Global(COUNT_i,COUNT_j) +
Ks_Global(COUNT_i,COUNT_j)
  END DO
 END DO

END SUBROUTINE Global_Stiff_Matrix_MARNON


SUBROUTINE Global_Stiff_Matrix_GeoNon
!-------------------------------------------------------------------------------
! The subroutine Global_Stiff_Matrix_GeoNon
! generates global stiffness matrix.
!-------------------------------------------------------------------------------
 USE MODULE_NAME
 IMPLICIT NONE
 INTEGER :: COUNT_DIMEN, COUNT_i, COUNT_j, locate

 ALLOCATE(K_Large_Global(DIMENSION_GLOBAL,DIMENSION_GLOBAL))
 ALLOCATE(K_Sigma_Global(DIMENSION_GLOBAL,DIMENSION_GLOBAL))

 locate = 0
 K_Large_Global = 0
 K_Sigma_Global = 0
 DO COUNT_DIMEN = 1, Num_Element_Stru
  CALL STIFF_MATRIX_GEONON(COUNT_DIMEN)
  CALL INITIAL_STRESS_MATRIX(COUNT_DIMEN)
  DO COUNT_i = 1, 4
```

```
   DO COUNT_j = 1, 4
    K_Large_Global(COUNT_i+locate,COUNT_j+locate) = K_Large_Global
(COUNT_i+locate,COUNT_j+locate) + K_Large(COUNT_i,COUNT_j)
    K_Sigma_Global(COUNT_i+locate,COUNT_j+locate) = K_Sigma_Global
(COUNT_i+locate,COUNT_j+locate) + K_Sigma(COUNT_i,COUNT_j)
   END DO
  END DO
  locate = locate + 2
 END DO

 DO COUNT_i = 1,DIMENSION_GLOBAL
  DO COUNT_j = 1,DIMENSION_GLOBAL
   K_Global(COUNT_i,COUNT_j) = K_Global(COUNT_i,COUNT_j)+
K_Large_Global(COUNT_i,COUNT_j)+ K_Sigma_Global(COUNT_i,
COUNT_j)
  END DO
 END DO

END SUBROUTINE Global_Stiff_Matrix_GeoNon


SUBROUTINE STIFF_MATRIX_MARNON(ELEMENT)
!-----------------------------------------------------------------------------
! The subroutine STIFF_MATRIX_MARNON generates
! element stiffness matrix.
!-----------------------------------------------------------------------------
 USE MODULE_NAME
 IMPLICIT NONE
 REAL, ALLOCATABLE, DIMENSION(:) :: u, w
 INTEGER :: COUNT, ELEMENT
 INTEGER :: i,j,COUNT_Gauss

 CALL Gauss_Beam(u,w)

 Kb = 0
 Ks = 0
 DO i = 1,4
  DO j = 1,4
   COUNT = Num_Gauss_Point*(ELEMENT-1)
   DO COUNT_Gauss = 1, Num_Gauss_Point
    COUNT = COUNT + 1
    x = 0.5*L*u(COUNT_Gauss) + 0.5*L
    CALL Shape_Function_Nonlinear(COUNT)
    Kb1  =  MATMUL(TRANSPOSE(Shape_Bb),Shape_Bb)*  Bending_Stiff_MAR
(COUNT)
    Kb(i,j) = Kb(i,j) + 0.5*L*w(COUNT_Gauss)*Kb1(i,j)
    Ks1=MATMUL(TRANSPOSE(Shape_Bs),Shape_Bs)*Shear_Stiff_MAR(COUNT)
    Ks(i,j) = Ks(i,j) + 0.5*L*w(COUNT_Gauss)*Ks1(i,j)
   END DO
```

```
  END DO
 END DO

END SUBROUTINE STIFF_MATRIX_MARNON


SUBROUTINE STIFF_MATRIX_GEONON(ELEMENT)
!-----------------------------------------------------------------------------
! The subroutine STIFF_MATRIX_GEONON generates
! element stiffness matrix for geometric nonlinearity.
!-----------------------------------------------------------------------------
 USE MODULE_NAME
 IMPLICIT NONE
 REAL, ALLOCATABLE, DIMENSION(:) :: u, w
 INTEGER :: i, j, ii, COUNT_Gauss, COUNT
 INTEGER :: ELEMENT

 CALL Gauss_Beam(u,w)

 K_b_lar = 0
 K_lar_b = 0
 K_lar_lar = 0
 DO i = 1,4
  DO j = 1,4
   COUNT = Num_Gauss_Point*(ELEMENT-1)
   DO COUNT_Gauss = 1, Num_Gauss_Point
    COUNT = COUNT + 1
    x = 0.5*L*u(COUNT_Gauss) + 0.5*L

    CALL Shape_Function_Nonlinear(COUNT)
    ALLOCATE(Shape_Bl_ELE(1,Dimension_Ele))

    DO ii = 1, Dimension_Ele
     Shape_Bl_ELE(1,ii) = Shape_Bl(ELEMENT,ii)
    END DO

    K1_b_lar = MATMUL(TRANSPOSE(Shape_Bb),Shape_Bl_ELE)* Dxb_STIFF
(COUNT)
    K_b_lar(i,j)=K_b_lar(i,j)+0.5*L*w(COUNT_Gauss)*K1_b_lar(i,j)
    K1_lar_b = MATMUL(TRANSPOSE(Shape_Bl_ELE),Shape_Bb)* Dxb_STIFF
(COUNT)
    K_lar_b(i,j) = K_lar_b(i,j) + 0.5*L*w(COUNT_Gauss)*K1_lar_b(i,j)
    K1_lar_lar = MATMUL(TRANSPOSE(Shape_Bl_ELE),Shape_Bl_ELE)*
Dxl_STIFF(COUNT)
    K_lar_lar(i,j) = K_lar_lar(i,j) + 0.5*L*w(COUNT_Gauss)*K1_lar_
lar(i,j)

    DEALLOCATE(Shape_Bl,Shape_Bl_ELE)

   END DO
  END DO
 END DO

 K_Large = K_b_lar + K_lar_b + K_lar_lar

END SUBROUTINE STIFF_MATRIX_GEONON
```

```
SUBROUTINE INITIAL_STRESS_MATRIX(ELEMENT)
!------------------------------------------------------------------------------
! The subroutine INITIAL_STRESS_MATRIX generates
! element initial stiffness matrix.
!------------------------------------------------------------------------------
 USE MODULE_NAME
 IMPLICIT NONE
 REAL, ALLOCATABLE, DIMENSION(:) :: u, w
 INTEGER :: i, ii, j, COUNT_Gauss, COUNT_LAYER, COUNT_MAR, MAX,
COUNT_E, COUNT
 INTEGER :: ELEMENT, mm

 CALL Gauss_Beam(u,w)

 K_Sigma = 0
 DO i = 1,4
  DO j = 1,4
   COUNT = Num_Gauss_Point*(ELEMENT-1)
   DO COUNT_Gauss = 1, Num_Gauss_Point
    COUNT = COUNT + 1
    x = 0.5*L*u(COUNT_Gauss) + 0.5*L

    CALL Shape_Function_Nonlinear(COUNT)

    K1_Sigma = MATMUL(TRANSPOSE(C),C)* N_STRESS(COUNT)
    K_Sigma(i,j)=K_Sigma(i,j)+0.5*L*w(COUNT_Gauss)*K1_Sigma(i,j)

    DEALLOCATE(Shape_Bl)

   END DO
  END DO
 END DO

END SUBROUTINE INITIAL_STRESS_MATRIX


SUBROUTINE Shape_Function_MARNON(GAUSSPOINT)
!------------------------------------------------------------------------------
! The subroutine Shape_Function_MARNON provides
! shape function for the current Gauss Point
!------------------------------------------------------------------------------
 USE MODULE_NAME
 IMPLICIT NONE
 INTEGER :: GAUSSPOINT

 Qb = Bending_Stiff_MAR(GAUSSPOINT)
 Qs = Shear_Stiff_MAR(GAUSSPOINT)

 Miu = 1/(1+12*Qb/(Qs*L**2))
 Shape_Bb(1,1) = 6*Miu*(1/L-2*x/L**2)/L
 Shape_Bb(1,2) = (3*Miu+1)/L-6*Miu*x/L**2
 Shape_Bb(1,3) = -6*Miu*(1/L-2*x/L**2)/L
```

```
 Shape_Bb(1,4) = -(1-3*Miu)/L-6*Miu*x/L**2
 Shape_Bs(1,1) = -1/L+Miu/L
 Shape_Bs(1,2) = Miu/2-1.0/2
 Shape_Bs(1,3) = 1/L-Miu/L
 Shape_Bs(1,4) = -1.0/2+Miu/2

END SUBROUTINE Shape_Function_MARNON


SUBROUTINE Shape_Function_Nonlinear(GAUSSPOINT)
!-------------------------------------------------------------------------
! The subroutine Shape_Function_Nonlinear finds
! nonlinear strain matrix Bl
!-------------------------------------------------------------------------
 USE MODULE_NAME
 IMPLICIT NONE
 INTEGER :: GAUSSPOINT, COUNT_ELE, m, i
 REAL, DIMENSION(1,1) :: Shape_Bl_1

 Qb = Bending_Stiff_MAR(GAUSSPOINT)
 Qs = Shear_Stiff_MAR(GAUSSPOINT)

 Miu = 1/(1+12*Qb/(Qs*L**2))
 C(1,1) = -1/L + Miu*(1/L - 6*x/L**2 + 6*x**2/L**3)
 C(1,2) = ((1/L-2*x/L**2) + Miu*(1/L-6*x/L**2+6*x**2/L**3))*L/2
 C(1,3) = 1/L + Miu*(6*x/L**2-1/L-6*x**2/L**3)
 C(1,4) = ((2*x/L**2-1/L) + Miu*(1/L-6*x/L**2+6*x**2/L**3))*L/2

 ALLOCATE(Shape_Bl(Num_Element_Stru,Dimension_Ele))
 DO COUNT_ELE = 1, Num_Element_Stru
  m = COUNT_ELE * 2 - 1
  DO i = 1, Dimension_Ele
   Displacement_Vec_Ele(i,1)=Displacement_Global(m,1)
   m = m + 1
  END DO

  DO i = 1, Dimension_Ele
   Shape_Bl_1 = MATMUL(C,Displacement_Vec_Ele)
   Shape_Bl(COUNT_ELE,i) = Shape_Bl_1(1,1) * C(1,i)
  END DO
 END DO

END SUBROUTINE Shape_Function_Nonlinear


SUBROUTINE EFF_INERTIA
!-------------------------------------------------------------------------
! The subroutine EFF_INERTIA finds position of
! neutral axis.
!-------------------------------------------------------------------------
 USE MODULE_NAME
 IMPLICIT NONE
 INTEGER :: i, j, k, COUNT, COUNT_STEEL
```

```
 ALLOCATE(FEy_CRACK(Num_Gauss_Point*Num_Element_Stru))
 ALLOCATE(FE_CRACK(Num_Gauss_Point*Num_Element_Stru))

 AXIS_CRACK = 0
 FEy_CRACK = 0
 FE_CRACK = 0
 DO i = 1, Num_Gauss_Point*Num_Element_Stru
  DO j = 1, Num_Material_Ele
   COUNT = Num_Layer_Material(j)*(i-1)
   DO k = 1, Num_Layer_Material(j)
    COUNT = COUNT + 1
    FEy_CRACK(i) = FEy_CRACK(i) + Db_MAR_CRACK(j,COUNT)*breadth*
((0.5*Thickness-Material_Bot(j,k))**2-(0.5*Thickness-Material_Top
(j,k))**2)
    FE_CRACK(i)  =  FE_CRACK(i)  +  2*Db_MAR_CRACK(j,COUNT)*breadth*
(Material_Top(j,k)-Material_Bot(j,k))
   END DO
  END DO

  AXIS_CRACK(i) = FEy_CRACK(i)/FE_CRACK(i)

 END DO

END SUBROUTINE EFF_INERTIA


SUBROUTINE MIGS (A,N,X,INDX)
!-----------------------------------------------------------------------------
! The subroutine MIGS finds inverse matrix.
!-----------------------------------------------------------------------------
 IMPLICIT NONE
 INTEGER, INTENT (IN) :: N
 INTEGER :: I,J,K
 INTEGER, INTENT (OUT), DIMENSION (N) :: INDX
 REAL, DIMENSION (N,N):: A
 REAL, INTENT (OUT), DIMENSION (N,N):: X
 REAL, DIMENSION (N,N) :: B

 DO I = 1, N
  DO J = 1, N
   B(I,J) = 0.0
  END DO
 END DO
 DO I = 1, N
  B(I,I) = 1.0
 END DO

 CALL ELGS (A,N,INDX)

 DO I = 1, N-1
  DO J = I+1, N
```

```
    DO K = 1, N
     B(INDX(J),K) = B(INDX(J),K)-A(INDX(J),I)*B(INDX(I),K)
    END DO
   END DO
  END DO

  DO I = 1, N
   X(N,I) = B(INDX(N),I)/A(INDX(N),N)
   DO J = N-1, 1, -1
    X(J,I) = B(INDX(J),I)
    DO K = J+1, N
     X(J,I) = X(J,I)-A(INDX(J),K)*X(K,I)
    END DO
    X(J,I) = X(J,I)/A(INDX(J),J)
   END DO
  END DO

END SUBROUTINE MIGS


SUBROUTINE ELGS (A,N,INDX)
  IMPLICIT NONE
  INTEGER, INTENT (IN) :: N
  INTEGER :: I,J,K,ITMP
  INTEGER, INTENT (OUT), DIMENSION (N) :: INDX
  REAL :: C1,PI,PI1,PJ
  REAL, INTENT (INOUT), DIMENSION (N,N) :: A
  REAL, DIMENSION (N) :: C

  ! Initialize the index
  DO I = 1, N
   INDX(I) = I
  END DO

  ! Find the rescaling factors, one from each row
  DO I = 1, N
   C1= 0.0
   DO J = 1, N
    C1 = AMAX1(C1,ABS(A(I,J)))
   END DO
   C(I) = C1
  END DO

  ! Search the pivoting (largest) element from each column
  DO J = 1, N-1
   PI1 = 0.0
   DO I = J, N
    PI = ABS(A(INDX(I),J))/C(INDX(I))
    IF (PI.GT.PI1) THEN
     PI1 = PI
```

```
    K = I
   ENDIF
  END DO

  ! Interchange the rows via INDX(N) to record pivoting order
  ITMP  = INDX(J)
  INDX(J) = INDX(K)
  INDX(K) = ITMP
  DO I = J+1, N
   PJ = A(INDX(I),J)/A(INDX(J),J)
   A(INDX(I),J) = PJ
   DO K = J+1, N
    A(INDX(I),K) = A(INDX(I),K)-PJ*A(INDX(J),K)
   END DO
  END DO
 END DO

END SUBROUTINE ELGS
```

Appendix H: Finite element code for composite beam element: Nonlinear analysis with bond–slip

This appendix provides the principal finite element code for nonlinear analysis of reinforced concrete beams with the consideration of bond–slip effect. In this part, only subroutines related to bond–slip effect are presented. The main program and the subroutines for concrete beam introduced in the previous sections are omitted herein.

```
SUBROUTINE Global_Stiff_Matrix_Linear
!--------------------------------------------------------------------------------
! The subroutine Global_Stiff_Matrix_Linear
! generates global stiffness matrix.
!--------------------------------------------------------------------------------
 USE MODULE_NAME
 IMPLICIT NONE
 INTEGER :: COUNT_DIMEN, COUNT_i, COUNT_j, locate, locate_bar, i, j

 CALL STIFF_MATRIX_LINEAR

 ALLOCATE(K_Global(DIMENSION_GLOBAL,DIMENSION_GLOBAL))

 locate = 0
 locate_bar = 0
 K_Global = 0
 DO COUNT_DIMEN = 1, NUM_ELEMENT_STRUC
  DO COUNT_i = 1, DOF_ELE
   i = COUNT_i + locate
   IF (COUNT_i > 4) THEN
    i = COUNT_i + (NUM_ELEMENT_STRUC-1)*2 + locate_bar
   END IF

   IF (COUNT_i > 6) THEN
    i = i + (1+INT((COUNT_i-7)/2))*(NUM_ELEMENT_STRUC-1)
   END IF

   DO COUNT_j = 1, DOF_ELE
    j = COUNT_j + locate
    IF (COUNT_j > 4) THEN
     j = COUNT_j + (NUM_ELEMENT_STRUC-1)*2 + locate_bar
    END IF
```

```
    IF (COUNT_j > 6) THEN
     j = j + (1+INT((COUNT_j-7)/2))*(NUM_ELEMENT_STRUC-1)
    END IF

    K_Global(i,j) = K_Global(i,j) + K0(COUNT_i,COUNT_j)

   END DO
  END DO
  locate = locate + 2
  locate_bar = locate_bar + 1
 END DO

END SUBROUTINE Global_Stiff_Matrix_Linear


SUBROUTINE STIFF_MATRIX_LINEAR
!------------------------------------------------------------------------------
! The subroutine STIFF_MATRIX_LINEAR generates
! stiffness matrix at linear elastic stage.
!------------------------------------------------------------------------------
 USE MODULE_NAME
 IMPLICIT NONE
 INTEGER :: i, j, COUNT_Gauss, COUNT_BAR
 INTEGER :: SHAPE_BAR1, SHAPE_BAR2
 REAL, ALLOCATABLE, DIMENSION(:) :: u, w

 {The element stiffness matrix for concrete beam has been discussed
in Appendix F. Omitted herein}

 ! For reinforcement with bond-slip
 CALL Shape_Function_Linear

 K_BAR = 0
 K_BOND = 0
 DO COUNT_BAR = 1, NUM_BAR_ELE
  SHAPE_BAR = 0

  SHAPE_BAR1 = COUNT_BAR*2 + 3
  SHAPE_BAR2 = COUNT_BAR*2 + 4

  SHAPE_BAR(1,SHAPE_BAR1) = BAR_1
  SHAPE_BAR(1,SHAPE_BAR2) = BAR_2

  DO i = 1, DOF_ELE
   DO j = 1, DOF_ELE
    DO COUNT_Gauss = 1, NUM_GAUSS_POINT
     K_BAR1=MATMUL(TRANSPOSE(SHAPE_BAR),SHAPE_BAR)* BAR_STIFF
(COUNT_BAR+1)
     K_BAR(i,j) = K_BAR(i,j) + 0.5*L*w(COUNT_Gauss)*K_BAR1(i,j)
     K_BOND1 = MATMUL(TRANSPOSE(SHAPE_BAR),SHAPE_BAR)* BOND_STIFF
(COUNT_BAR+1,COUNT_Gauss)
     K_BOND(i,j) = K_BOND(i,j) + 0.5*L*w(COUNT_Gauss)*K_BOND1(i,j)
```

```
    END DO
   END DO
  END DO
 END DO

 K0 = Kb + Ks + K_BAR + K_BOND

END SUBROUTINE STIFF_MATRIX_LINEAR


SUBROUTINE ELE_STIFF_LINEAR
!--------------------------------------------------------------------------------
! The subroutine ELE_STIFF_LINEAR generates
! element stiffness at linear elastic stage.
!--------------------------------------------------------------------------------
 USE MODULE_NAME
 IMPLICIT NONE
 INTEGER :: COUNT_MAT, i, ilayer
 INTEGER :: WL
 REAL, ALLOCATABLE, DIMENSION(:) :: u, w

 {The element stiffness for concrete beam has been discussed in
Appendix F. Omitted herein}

 ! For reinforcement with bond-slip
 IF (NUM_MATERIAL_ELE > 1) THEN
  DO COUNT_MAT = 2, NUM_MATERIAL_ELE
   BAR_STIFF(COUNT_MAT) = 0
   DO ilayer = 1, NUM_LAYER_MATERIAL(COUNT_MAT)
    BAR_STIFF(COUNT_MAT) = BAR_STIFF(COUNT_MAT) + BREADTH*Db
(COUNT_MAT)*(Material_Top(COUNT_MAT,ilayer) - Material_Bot
(COUNT_MAT,ilayer))
   END DO
  END DO

  CALL Gauss_Beam(u,w)

  DO COUNT_MAT = 2, NUM_MATERIAL_ELE
   DO i = 1, NUM_GAUSS_POINT
    WL = i
    BOND_STIFF(COUNT_MAT,i) = 0
    DO j = 1, NUM_LAYER_MATERIAL(COUNT_MAT)
     BOND_STIFF(COUNT_MAT,i)=BOND_STIFF(COUNT_MAT,i)+D_bond
(COUNT_MAT)*PERIMETER(COUNT_MAT)*L*w(WL)
    END DO
   END DO
  END DO
 END IF

END SUBROUTINE ELE_STIFF_LINEAR
```

```
SUBROUTINE Shape_Function_Linear
!-------------------------------------------------------------------------------
! The subroutine Shape_Function_Linear provides
! shape function for beam element.
!-------------------------------------------------------------------------------
 USE MODULE_NAME
 IMPLICIT NONE

 {The shape function for concrete beam has been discussed in Appendix F.
Omitted herein}

 ! For reinforcement
 BAR_1 = -1.0/L
 BAR_2 = 1.0/L

END SUBROUTINE Shape_Function_Linear


SUBROUTINE MARNON_STRAIN
!-------------------------------------------------------------------------------
! The subroutine MARNON_STRAIN finds strains
! at Gauss Point.
!-------------------------------------------------------------------------------
 USE MODULE_NAME
 IMPLICIT NONE
 REAL :: ELE_X, a
 INTEGER :: i, j, COUNT_LAYER, COUNT_Gauss,COUNT_DIMN, COUNT,
COUNT_BAR
 INTEGER :: COUNT_SHAPE, COUNT_SS, COUNT_ELE, m, n
 INTEGER :: SHAPE_BAR1, SHAPE_BAR2, WL
 REAL, ALLOCATABLE, DIMENSION(:) :: u, w

 ALLOCATE(STRAIN_X(NUM_MATERIAL_ELE,DIMN_STRAIN))
 ALLOCATE(STRAIN_Z(NUM_MATERIAL_ELE,DIMN_STRAIN))
 ALLOCATE(LOCATE_Z(NUM_MATERIAL_ELE,DIMN_STRAIN))

 CALL Gauss_Beam(u,w)

 DO i = 1, NUM_MATERIAL_ELE
  DO j = 1, NUM_ELEMENT_STRUC
   DO COUNT_Gauss = 1, NUM_GAUSS_POINT
    DO COUNT_LAYER = 1, NUM_LAYER_MATERIAL(i)
     COUNT_DIMN = NUM_GAUSS_POINT*NUM_LAYER_MATERIAL(i)*(j-1) +
NUM_LAYER_MATERIAL(i)*(COUNT_Gauss-1) + COUNT_LAYER
     ELE_X = 0.5*L*u(COUNT_Gauss) + 0.5*L
     STRAIN_X(i,COUNT_DIMN) = ELE_X + (j-1)*L
     STRAIN_Z(i,COUNT_DIMN) = 0.5 * (Material_Bot(i,COUNT_LAYER)+
Material_Top(i,COUNT_LAYER))
    END DO
   END DO
```

```
  END DO
 END DO

 COUNT = 0
 INCREMENT_Bl_STRAIN = 0
 INCREMENT_Bb_STRAIN = 0
 DO i = 1, NUM_MATERIAL_ELE
  COUNT_SHAPE = 0
  COUNT_SS = 0
  COUNT_ELE = 0
  DO j = 1, NUM_GAUSS_POINT*NUM_LAYER_MATERIAL(i)*NUM_ELEMENT_STRUC
   COUNT = COUNT + 1
   COUNT_SS = COUNT_SS + 1
   COUNT_SHAPE = (COUNT_SS-1)/NUM_LAYER_MATERIAL(i) + 1
   COUNT_ELE = (COUNT_SS-1)/(NUM_LAYER_MATERIAL(i)*NUM_GAUSS_POINT)+1
   WL = (COUNT_SS-1)/NUM_LAYER_MATERIAL(i)+1-(COUNT_ELE-1)*
NUM_GAUSS_POINT

   m = (INT(STRAIN_X(i,j)/L) + 1) * 2 - 1
   DO n = 1, Dimension_Ele
    Displacement_Vec_Ele(n,1) = Displacement_Global(m,1)
    Increment_Vec_Ele(n,1) = Increment_Global(m,1)

    IF (n <= 5) THEN
     m = m + 1
    END IF

    IF (n == 4) THEN
     m = m + (NUM_ELEMENT_STRUC-COUNT_ELE)*2 + (COUNT_ELE-1)
    END IF

    IF (n > 5) THEN
     IF (n/2 == n/2.0) THEN
      a = 1
     ELSE
      a = 0
     END IF
     m = m + (NUM_ELEMENT_STRUC-1)*a + 1
    END IF
   END DO

   x = STRAIN_X(i,j) - INT(STRAIN_X(i,j)/L) * L

   IF (STEP_NON == 1) THEN
    IF (COUNT_ITER == 0) THEN
     CALL Shape_Function_Linear
     CALL Shape_Function_Nonlinear(COUNT_SHAPE)

     SHAPE_BAR = 0
     COUNT_BAR = i - 1

     SHAPE_BAR1 = COUNT_BAR*2 + 3
     SHAPE_BAR2 = COUNT_BAR*2 + 4
```

```
    SHAPE_BAR(1,SHAPE_BAR1) = BAR_1
    SHAPE_BAR(1,SHAPE_BAR2) = BAR_2

    DO n = 1, Dimension_Ele
      Bending_Strain_CHECK(i,j) = Bending_Strain_CHECK(i,j) +
Shape_Bb(1,n) * Displacement_Vec_Ele(n,1)
      Shear_Strain_CHECK(i,j) = Shear_Strain_CHECK(i,j) + Shape_Bs
(1,n) * Displacement_Vec_Ele(n,1)
      Large_Strain(i,j) = 0.5 * Shape_Bl(COUNT_ELE,n) *
Displacement_Vec_Ele(n,1)
    END DO

    IF (NEU_AXIS == 0.5*Thickness) THEN
     IF (Bending_Strain_CHECK(i,j) > 0) THEN
      STRAIN_Z(i,j) = -STRAIN_Z(i,j)
     END IF

     DO n = 1, Dimension_Ele
      Bending_Strain_CHECK_Z(i,j) = Bending_Strain_CHECK_Z(i,j) +
(STRAIN_Z(i,j) * Shape_Bb(1,n) + 0.5 * Shape_Bl(COUNT_ELE,n)) *
Displacement_Vec_Ele(n,1)
     END DO
    END IF

    IF (NEU_AXIS /= 0.5*Thickness) THEN
     LOCATE_Z(i,j) = 0.5*Thickness-STRAIN_Z(i,j)-NEU_AXIS

     IF (Bending_Strain_CHECK(i,j) < 0) THEN
      LOCATE_Z(i,j) = - LOCATE_Z(i,j)
     END IF

     DO n = 1, Dimension_Ele
      Bending_Strain_CHECK_Z(i,j) = Bending_Strain_CHECK_Z(i,j) +
(LOCATE_Z(i,j) * Shape_Bb(1,n) + 0.5 * Shape_Bl(COUNT_ELE,n)) *
Displacement_Vec_Ele(n,1)
     END DO
    END IF

    IF (i > 1) THEN
     Bar_Strain_CHECK(i,j) = Bending_Strain_CHECK_Z(i,j)
     Bond_Strain_CHECK(i,j) = Bar_Strain_CHECK(i,j) -
Bending_Strain_CHECK_Z(i,j)
    END IF

   ELSE
    CALL Shape_Function_MARNON(COUNT_SHAPE)
    CALL Shape_Function_Nonlinear(COUNT_SHAPE)

    SHAPE_BAR = 0
    COUNT_BAR = i - 1

    SHAPE_BAR1 = COUNT_BAR*2 + 3
    SHAPE_BAR2 = COUNT_BAR*2 + 4
```

```
    SHAPE_BAR(1,SHAPE_BAR1) = BAR_1
    SHAPE_BAR(1,SHAPE_BAR2) = BAR_2

    DO n = 1, Dimension_Ele
     Bending_Strain_CHECK(i,j)=Bending_Strain_CHECK(i,j)+
Shape_Bb(1,n) * Increment_Vec_Ele(n,1)
     Shear_Strain_CHECK(i,j) = Shear_Strain_CHECK(i,j) +
Shape_Bs(1,n) * Increment_Vec_Ele(n,1)
     INCREMENT_Bl_STRAIN(i,j)=INCREMENT_Bl_STRAIN(i,j) +
Shape_Bl(COUNT_ELE,n) * Increment_Vec_Ele(n,1)
     INCREMENT_Bb_STRAIN(i,j)=INCREMENT_Bb_STRAIN(i,j) +
Shape_Bb(1,n) * Increment_Vec_Ele(n,1)
    END DO

    IF (AXIS_CRACK(COUNT_SHAPE) == 0.5*Thickness) THEN
     IF (Bending_Strain_CHECK(i,j) > 0) THEN
      STRAIN_Z(i,j) = -STRAIN_Z(i,j)
     END IF

     DO n = 1, Dimension_Ele
      Bending_Strain_CHECK_Z(i,j) = Bending_Strain_CHECK_Z(i,j) +
(STRAIN_Z(i,j) * Shape_Bb(1,n) + Shape_Bl(COUNT_ELE,n)) *
Increment_Vec_Ele(n,1)
     END DO
    END IF

    IF (AXIS_CRACK(COUNT_SHAPE) /= 0.5*Thickness) THEN
     LOCATE_Z(i,j)=0.5*Thickness-STRAIN_Z(i,j)-AXIS_CRACK
(COUNT_SHAPE)

     IF (Bending_Strain_CHECK(i,j) < 0) THEN
      LOCATE_Z(i,j) = - LOCATE_Z(i,j)
     END IF

     DO n = 1, Dimension_Ele
      Bending_Strain_CHECK_Z(i,j) = Bending_Strain_CHECK_Z(i,j) +
(LOCATE_Z(i,j) * Shape_Bb(1,n) + Shape_Bl(COUNT_ELE,n)) *
Increment_Vec_Ele(n,1)
     END DO
    END IF

    IF (i > 1) THEN
     DO n = 1, Dimension_Ele
      Bar_Strain_CHECK(i,j)=Bar_Strain_CHECK(i,j)+SHAPE_BAR(1,n)*
Increment_Vec_Ele(n,1)
     END DO
     Bond_Strain_CHECK(i,j)=Bar_Strain_CHECK(i,j) -
Bending_Strain_CHECK_Z(i,j)
    END IF
   END IF
  END IF
```

```
  IF (STEP_NON > 1) THEN
   CALL Shape_Function_MARNON(COUNT_SHAPE)
   CALL Shape_Function_Nonlinear(COUNT_SHAPE)

   SHAPE_BAR = 0
   COUNT_BAR = i - 1

   SHAPE_BAR1 = COUNT_BAR*2 + 3
   SHAPE_BAR2 = COUNT_BAR*2 + 4

   SHAPE_BAR(1,SHAPE_BAR1) = BAR_1
   SHAPE_BAR(1,SHAPE_BAR2) = BAR_2

   DO n = 1, Dimension_Ele
    Bending_Strain_CHECK(i,j)=Bending_Strain_CHECK(i,j)+
Shape_Bb(1,n) * Increment_Vec_Ele(n,1)
    Shear_Strain_CHECK(i,j) = Shear_Strain_CHECK(i,j) + Shape_Bs
(1,n) * Increment_Vec_Ele(n,1)
    INCREMENT_Bl_STRAIN(i,j) = INCREMENT_Bl_STRAIN(i,j) + Shape_Bl
(COUNT_ELE,n) * Increment_Vec_Ele(n,1)
    INCREMENT_Bb_STRAIN(i,j) = INCREMENT_Bb_STRAIN(i,j) + Shape_Bb
(1,n) * Increment_Vec_Ele(n,1)
   END DO

   IF (AXIS_CRACK(COUNT_SHAPE) == 0.5*Thickness) THEN
    IF (Bending_Strain_CHECK(i,j) > 0) THEN
     STRAIN_Z(i,j) = -STRAIN_Z(i,j)
    END IF

    DO n = 1, Dimension_Ele
     Bending_Strain_CHECK_Z(i,j) = Bending_Strain_CHECK_Z(i,j) +
(STRAIN_Z(i,j) * Shape_Bb(1,n) + Shape_Bl(COUNT_ELE,n)) *
Increment_Vec_Ele(n,1)
    END DO
   END IF

   IF (AXIS_CRACK(COUNT_SHAPE) /= 0.5*Thickness) THEN
    LOCATE_Z(i,j)=0.5*Thickness-STRAIN_Z(i,j)-AXIS_CRACK
(COUNT_SHAPE)

    IF (Bending_Strain_CHECK(i,j) < 0) THEN
     LOCATE_Z(i,j)=- LOCATE_Z(i,j)
    END IF

    DO n = 1, Dimension_Ele
     Bending_Strain_CHECK_Z(i,j)=Bending_Strain_CHECK_Z(i,j)+
(LOCATE_Z(i,j)*Shape_Bb(1,n)+Shape_Bl(COUNT_ELE,n))*
Increment_Vec_Ele(n,1)
    END DO
   END IF

   IF (i > 1) THEN
    DO n = 1, Dimension_Ele
```

```
     Bar_Strain_CHECK(i,j)=Bar_Strain_CHECK(i,j)+SHAPE_BAR(1,n)*
Increment_Vec_Ele(n,1)
     END DO
     Bond_Strain_CHECK(i,j)=Bar_Strain_CHECK(i,j)-
Bending_Strain_CHECK_Z(i,j)
    END IF
   END IF

   IF (Bending_Strain_CHECK_Z(i,j) >= 0) THEN
    Bending_Strain_PRINCIPAL(i,j) = Bending_Strain_CHECK_Z(i,j)/2 +
SQRT((Bending_Strain_CHECK_Z(i,j)/2)**2+(Shear_Strain_CHECK(i,
j)/2)**2)
   ELSE
    Bending_Strain_PRINCIPAL(i,j) = Bending_Strain_CHECK_Z(i,j)/2 -
SQRT((Bending_Strain_CHECK_Z(i,j)/2)**2 + (Shear_Strain_CHECK
(i,j)/2)**2)
   END IF

   DEALLOCATE(Shape_Bl)
  END DO
 END DO

END SUBROUTINE MARNON_STRAIN


SUBROUTINE INTERNAL_FORCE
!-------------------------------------------------------------------------------
! The subroutine INTERNAL_FORCE finds element
! internal force.
!-------------------------------------------------------------------------------
 USE MODULE_NAME
 IMPLICIT NONE
 REAL :: DISTANCE_Y
 INTEGER :: COUNT, COUNT_GAUSS, i, j, locate, locate_bar, locate_j,
imaterial, iDIMN
 INTEGER :: COUNT_BAR, WL
 INTEGER :: SHAPE_BAR1, SHAPE_BAR2
 REAL, ALLOCATABLE, DIMENSION(:) :: u, w

 ALLOCATE(INTERNAL_FORCE_ELE(NUM_ELEMENT_STRUC,Dimension_Ele))
 ALLOCATE(INTERNAL_FORCE_ELE_1(Dimension_Ele,1),
INTERNAL_FORCE_ELE_2(Dimension_Ele,1))
 ALLOCATE(INTERNAL_BAR_ELE(NUM_MATERIAL_ELE,Dimension_Ele))
 ALLOCATE(INTERNAL_BOND_ELE(NUM_MATERIAL_ELE,Dimension_Ele))

 CALL Gauss_Beam(u,w)

 IF (STEP_NON == 1 .AND. COUNT_ITER == 0) THEN
  DO COUNT_GAUSS = 1, NUM_GAUSS_POINT*NUM_ELEMENT_STRUC
   N_STRESS(COUNT_GAUSS) = 0
   M_STRESS(COUNT_GAUSS) = 0
```

```
   Q_STRESS(COUNT_GAUSS) = 0
   DO i = 1, NUM_MATERIAL_ELE
    COUNT = (COUNT_GAUSS-1)*NUM_LAYER_MATERIAL(i)
    DO j = 1, NUM_LAYER_MATERIAL(i)
     COUNT = COUNT + 1
     IF (NEU_AXIS == 0.5*Thickness) THEN
      DISTANCE_Y = 0.5*(Material_Top(i,j)+Material_Bot(i,j))
      IF (Bending_Strain_CHECK(i,COUNT) > 0) THEN
       DISTANCE_Y = -0.5*(Material_Top(i,j)+Material_Bot(i,j))
      END IF
     END IF

     IF (NEU_AXIS /= 0.5*Thickness) THEN
      DISTANCE_Y=0.5*Thickness-0.5*(Material_Top(i,j)+
Material_Bot(i,j))-NEU_AXIS
      IF (Bending_Strain_CHECK(i,COUNT) < 0) THEN
       DISTANCE_Y = - DISTANCE_Y
      END IF
     END IF

     N_STRESS(COUNT_GAUSS) = N_STRESS(COUNT_GAUSS) + BREADTH*Db(i)*
(Material_Top(i,j)-Material_Bot(i,j))*Bending_Strain_CHECK_Z
(i,COUNT)
     M_STRESS(COUNT_GAUSS) = M_STRESS(COUNT_GAUSS) + BREADTH*Db(i)*
(Material_Top(i,j)-Material_Bot(i,j))*Bending_Strain_CHECK_Z
(i,COUNT)*DISTANCE_Y
     Q_STRESS(COUNT_GAUSS) = Q_STRESS(COUNT_GAUSS) + BREADTH*Ds(i)*
(Material_Top(i,j)-Material_Bot(i,j))*Shear_Strain_CHECK
(i,COUNT)*5/6
    END DO

    IF (i > 1) THEN
     BAR_STRESS(i,COUNT_GAUSS) = 0
     BOND_STRESS(i,COUNT_GAUSS) = 0

     WL=COUNT_GAUSS - NUM_GAUSS_POINT*INT((COUNT_GAUSS-1)/
NUM_GAUSS_POINT)

     COUNT=(COUNT_GAUSS-1)*NUM_LAYER_MATERIAL(i)
     DO j = 1, NUM_LAYER_MATERIAL(i)
      COUNT = COUNT + 1

      IF (NEU_AXIS == 0.5*Thickness) THEN
       DISTANCE_Y = 0.5*(Material_Top(i,j)+Material_Bot(i,j))
       IF (Bending_Strain_CHECK(i,COUNT) > 0) THEN
        DISTANCE_Y = -0.5*(Material_Top(i,j)+Material_Bot(i,j))
       END IF
      END IF

      IF (NEU_AXIS /= 0.5*Thickness) THEN
       DISTANCE_Y=0.5*Thickness-0.5*(Material_Top(i,j)+
Material_Bot(i,j))-NEU_AXIS
```

```
      IF (Bending_Strain_CHECK(i,COUNT) < 0) THEN
       DISTANCE_Y = - DISTANCE_Y
      END IF
     END IF

     BAR_STRESS(i,COUNT_GAUSS)=BAR_STRESS(i,COUNT_GAUSS)+
BREADTH*Db(i)*(Material_Top(i,j)-Material_Bot(i,j))*
Bar_Strain_CHECK(i,COUNT)
     BOND_STRESS(i,COUNT_GAUSS)=BOND_STRESS(i,COUNT_GAUSS)+
PERIMETER(i)*D_bond(i)*Bond_Strain_CHECK(i,COUNT)*LENGTH*
w(WL)
    END DO
   END IF
  END DO
 END DO
ELSE
 DO COUNT_GAUSS = 1, NUM_GAUSS_POINT*NUM_ELEMENT_STRUC
  N_STRESS(COUNT_GAUSS) = 0
  M_STRESS(COUNT_GAUSS) = 0
  Q_STRESS(COUNT_GAUSS) = 0
  DO i = 1, NUM_MATERIAL_ELE
   COUNT = (COUNT_GAUSS-1)*NUM_LAYER_MATERIAL(i)
   DO j = 1, NUM_LAYER_MATERIAL(i)
    COUNT = COUNT + 1
    IF (AXIS_CRACK(COUNT_GAUSS) == 0.5*Thickness) THEN
     DISTANCE_Y = 0.5*(Material_Top(i,j)+Material_Bot(i,j))
     IF (Bending_Strain_CHECK(i,COUNT) > 0) THEN
      DISTANCE_Y = -0.5*(Material_Top(i,j)+Material_Bot(i,j))
     END IF
    END IF

    IF (AXIS_CRACK(COUNT_GAUSS) /= 0.5*Thickness) THEN
     DISTANCE_Y=0.5*Thickness-0.5*(Material_Top(i,j)+
Material_Bot(i,j))-AXIS_CRACK(COUNT_GAUSS)
     IF (Bending_Strain_CHECK(i,COUNT) < 0) THEN
      DISTANCE_Y = - DISTANCE_Y
     END IF
    END IF

    N_STRESS(COUNT_GAUSS)=N_STRESS(COUNT_GAUSS)+
BREADTH*Db_MAR_CRACK(i,COUNT)*(Material_Top(i,j)-Material_Bot
(i,j))*Bending_Strain_CHECK_Z(i,COUNT)
    M_STRESS(COUNT_GAUSS)=M_STRESS(COUNT_GAUSS)+
BREADTH*Db_MAR_CRACK(i,COUNT)*(Material_Top(i,j)-Material_Bot
(i,j))*Bending_Strain_CHECK_Z(i,COUNT)*DISTANCE_Y
    Q_STRESS(COUNT_GAUSS) = Q_STRESS(COUNT_GAUSS) + BREADTH*Ds_MAR
(i,COUNT)*(Material_Top(i,j)-Material_Bot(i,j))*
Shear_Strain_CHECK(i,COUNT)*5/6
   END DO
```

```
    IF (i > 1) THEN
     BAR_STRESS(i,COUNT_GAUSS) = 0
     BOND_STRESS(i,COUNT_GAUSS) = 0

     WL = COUNT_GAUSS - NUM_GAUSS_POINT*INT((COUNT_GAUSS-1)/
NUM_GAUSS_POINT)

     COUNT = (COUNT_GAUSS-1)*NUM_LAYER_MATERIAL(i)
     DO j = 1, NUM_LAYER_MATERIAL(i)
      COUNT = COUNT + 1
      IF (AXIS_CRACK(COUNT_GAUSS) == 0.5*Thickness) THEN
       DISTANCE_Y = 0.5*(Material_Top(i,j)+Material_Bot(i,j))
       IF (Bending_Strain_CHECK(i,COUNT) > 0) THEN
        DISTANCE_Y = -0.5*(Material_Top(i,j)+Material_Bot(i,j))
       END IF
      END IF

      IF (AXIS_CRACK(COUNT_GAUSS) /= 0.5*Thickness) THEN
       DISTANCE_Y = 0.5*Thickness-0.5*(Material_Top(i,j)+
Material_Bot(i,j))-AXIS_CRACK(COUNT_GAUSS)
       IF (Bending_Strain_CHECK(i,COUNT) < 0) THEN
        DISTANCE_Y = - DISTANCE_Y
       END IF
      END IF

      BAR_STRESS(i,COUNT_GAUSS) = BAR_STRESS(i,COUNT_GAUSS) +
BREADTH*Db_MAR(i,COUNT)*(Material_Top(i,j)-Material_Bot(i,j))*
Bar_Strain_CHECK(i,COUNT)
      BOND_STRESS(i,COUNT_GAUSS) = BOND_STRESS(i,COUNT_GAUSS) +
PERIMETER(i)*D_bond_MAR(i,COUNT)*Bond_Strain_CHECK(i,COUNT)*
LENGTH*w(WL)
     END DO
    END IF
   END DO
  END DO
 END IF

 INTERNAL_FORCE_ELE = 0
 DO i = 1, NUM_ELEMENT_STRUC
  DO j = 1, Dimension_Ele
   COUNT = NUM_GAUSS_POINT*(i-1)
   DO COUNT_GAUSS = 1, NUM_GAUSS_POINT
    COUNT = COUNT + 1
    x = 0.5*L*u(COUNT_Gauss) + 0.5*(x2+x1)

    IF (STEP_NON == 1 .AND. COUNT_ITER == 0) THEN
     CALL Shape_Function_Linear
    ELSE
     CALL Shape_Function_MARNON(COUNT)
    END IF

    INTERNAL_FORCE_ELE_1 = TRANSPOSE(Shape_Bb)* M_STRESS(COUNT)
```

```
    INTERNAL_FORCE_ELE_2 = TRANSPOSE(Shape_Bs)* Q_STRESS(COUNT)
    INTERNAL_FORCE_ELE(i,j) = INTERNAL_FORCE_ELE(i,j) + 0.5*L*
w(COUNT_Gauss)*(INTERNAL_FORCE_ELE_1(j,1)+INTERNAL_FORCE_ELE_2
(j,1))
   END DO
  END DO
 END DO

 INTERNAL_BAR_ELE = 0
 INTERNAL_BOND_ELE = 0
 IF (NUM_MATERIAL_ELE > 1) THEN
  DO imaterial = 2, NUM_MATERIAL_ELE
   SHAPE_BAR = 0
   COUNT_BAR = imaterial - 1

   SHAPE_BAR1 = COUNT_BAR*2 + 3
   SHAPE_BAR2 = COUNT_BAR*2 + 4

   DO i = 1, NUM_ELEMENT_STRUC
    COUNT = NUM_GAUSS_POINT*(i-1)
    DO COUNT_GAUSS = 1, NUM_GAUSS_POINT
     COUNT = COUNT + 1
     x = 0.5*L*u(COUNT_Gauss) + 0.5*L

     IF (STEP_NON == 1 .AND. COUNT_ITER == 0) THEN
      CALL Shape_Function_Linear
     ELSE
      CALL Shape_Function_MARNON(COUNT)
     END IF

     SHAPE_BAR(1,SHAPE_BAR1) = BAR_1
     SHAPE_BAR(1,SHAPE_BAR2) = BAR_2

     DO iDIMN = 1, Dimension_Ele
      INTERNAL_BAR_ELE(imaterial,iDIMN) = SHAPE_BAR(1,iDIMN)*
BAR_STRESS(imaterial,COUNT)
      INTERNAL_BOND_ELE(imaterial,iDIMN) = SHAPE_BAR(1,iDIMN)*
BOND_STRESS(imaterial,COUNT)
     END DO

     DO j = 1, Dimension_Ele
      INTERNAL_FORCE_ELE(i,j) = INTERNAL_FORCE_ELE(i,j) + 0.5*L*
w(COUNT_Gauss)*(INTERNAL_BOND_ELE(imaterial,j)+
INTERNAL_BAR_ELE(imaterial,j))
     END DO
    END DO
   END DO
  END DO
 END IF

 ALLOCATE(GLOBAL_INTER_FORCE(DIMENSION_GLOBAL,1))

 GLOBAL_INTER_FORCE = 0
 locate = 0
```

```
locate_bar = 0
DO i = 1, NUM_ELEMENT_STRUC
 DO j = 1, Dimension_Ele
  locate_j = j + locate
  IF (j > 4) THEN
   locate_j = j + (NUM_ELEMENT_STRUC-1)*2 + locate_bar
  END IF

  IF (j > 6) THEN
   locate_j = locate_j + (1+INT((j-7)/2))*(NUM_ELEMENT_STRUC-1)
  END IF

  GLOBAL_INTER_FORCE(locate_j,1)=GLOBAL_INTER_FORCE(locate_j,1)+
INTERNAL_FORCE_ELE(i,j)
 END DO
 locate = locate + 2
 locate_bar = locate_bar + 1
END DO

ALLOCATE(INTER_FORCE(Num_Unknown,1))

SKIP = 0
DO i = 1,DIMENSION_GLOBAL
 NONZERO = 0
 DO j = 1, R_0
  IF (i == ROW_ZERO(j)) THEN
   NONZERO = NONZERO + 1
  END IF
 END DO

 IF (NONZERO == 0) THEN
  SKIP = SKIP+1
  INTER_FORCE(SKIP,1) = GLOBAL_INTER_FORCE(i,1)
 END IF
END DO

END SUBROUTINE INTERNAL_FORCE


SUBROUTINE ELE_STIFF_MARNON
!------------------------------------------------------------------------------
! The subroutine ELE_STIFF_MARNON generates
! element stiffness for nonlinear analysis.
!------------------------------------------------------------------------------
 USE MODULE_NAME
 IMPLICIT NONE
 INTEGER :: i, j, k, m, n, COUNT, COUNT_LAYER, COUNT_GAUSS, COUNT_MAR, WL
 INTEGER :: COUNT_R, COUNT_C
 REAL :: STRAIN_1, STRAIN_2, EFF_HEIGHT, OPP_HEIGHT
 REAL, ALLOCATABLE, DIMENSION(:) :: u, w
```

```
 ALLOCATE(Dbb_MAR(NUM_MATERIAL_ELE,NUM_GAUSS_POINT*
NUM_ELEMENT_STRUC))
 ALLOCATE(Dss_MAR(NUM_MATERIAL_ELE,NUM_GAUSS_POINT*
NUM_ELEMENT_STRUC))
 ALLOCATE(Dxl_MAR(NUM_MATERIAL_ELE,NUM_GAUSS_POINT*
NUM_ELEMENT_STRUC))
 ALLOCATE(Dxb_MAR(NUM_MATERIAL_ELE,NUM_GAUSS_POINT*
NUM_ELEMENT_STRUC))
 ALLOCATE(BAR_STIFF_MAR(NUM_MATERIAL_ELE,NUM_GAUSS_POINT*
NUM_ELEMENT_STRUC))
 ALLOCATE(BOND_STIFF_MAR(NUM_MATERIAL_ELE,NUM_GAUSS_POINT*
NUM_ELEMENT_STRUC))

 {The nonlinear element stiffness for concrete beam has been
discussed in Appendix G. Omitted herein}

 ! For reinforcement
 DO i = 2, NUM_MATERIAL_ELE
  DO j =1, NUM_GAUSS_POINT*NUM_LAYER_MATERIAL(i)*NUM_ELEMENT_STRUC
   COUNT_GAUSS = INT((j-1)/NUM_LAYER_MATERIAL(1)) + 1
   WL = COUNT_GAUSS - NUM_GAUSS_POINT*INT((COUNT_GAUSS-1)/
NUM_GAUSS_POINT)

   IF (Crack_Stress(i) /= 0) THEN ! FRP
    Crack_Strain(i) = Crack_Stress(i)/Db(i)

    IF (Bar_Strain_CHECK(i,j) >= 0) THEN
     IF (Bar_Strain_CHECK(i,j) > Crack_Strain(i)) THEN
      Db_MAR(i,j) = 0
     ENDIF
    ENDIF

    IF (Bar_Strain_CHECK(i,j) < 0) THEN
     Db_MAR(i,j) = 0
    END IF

    ! FRP BOND
    SLIP = ABS(Bond_Strain_CHECK(i,j)*LENGTH*w(WL))
    IF (SLIP == 0) THEN
     D_bond_MAR(i,j) = D_bond(i)
    ELSE IF (SLIP <= s1_FRP) THEN
     D_bond_MAR(i,j) = ((SLIP/s1_FRP)**SHEAR_CO_FRP)*SHEAR_MAX_FRP/
SLIP
    ELSE IF (SLIP <= s3_FRP) THEN
     D_bond_MAR(i,j) = (1-SOFTENING*(SLIP/s1_FRP-1))*SHEAR_MAX_FRP/
SLIP
    ELSE
     D_bond_MAR(i,j) = SHEAR_FAIL_FRP/SLIP
    END IF
   END IF
```

```
    IF (Crack_Stress(i) == 0) THEN ! STEEL
     Compression_Strain(i) = Ulti_Stress(i)/Db(i)

     IF (Bar_Strain_CHECK(i,j) >= 0) THEN
      IF (Bar_Strain_CHECK(i,j) > Tension_Strain(i)) THEN
       Db_MAR(i,j) = Ulti_Stress(i)/Bar_Strain_CHECK(i,j)
      END IF
     END IF

     IF (Bar_Strain_CHECK(i,j) < 0) THEN
      IF (ABS(Bar_Strain_CHECK(i,j)) > Compression_Strain(i)) THEN
       Db_MAR(i,j) = Ulti_Stress(i)/ABS(Bar_Strain_CHECK(i,j))
      END IF
     END IF

     ! STEEL BOND
     SLIP = ABS(Bond_Strain_CHECK(i,j)*LENGTH*w(WL))

     IF (SLIP == 0) THEN
      D_bond_MAR(i,j) = D_bond(i)
     ELSE IF (SLIP <= s1) THEN
      D_bond_MAR(i,j) = (SHEAR_MAX*(SLIP/s1)**SHEAR_CO)/SLIP
     ELSE IF (SLIP <= s2) THEN
      D_bond_MAR(i,j) = SHEAR_MAX/SLIP
     ELSE IF (SLIP <= s3) THEN
      D_bond_MAR(i,j)=(SHEAR_MAX-(SHEAR_MAX-SHEAR_FAIL)*
((SLIP-s2)/(s3-s2)))/SLIP
     ELSE
      D_bond_MAR(i,j)=SHEAR_FAIL/SLIP
     END IF
    END IF
   END DO
  END DO

  DO COUNT_MAR = 2, NUM_MATERIAL_ELE
   COUNT = 0
   DO i = 1, NUM_GAUSS_POINT*NUM_ELEMENT_STRUC
    BOND_STIFF_MAR(COUNT_MAR,i) = 0
    BAR_STIFF_MAR(COUNT_MAR,i) = 0

    WL = i - INT((i-1)/NUM_GAUSS_POINT)*NUM_GAUSS_POINT

    DO j = 1, NUM_LAYER_MATERIAL(COUNT_MAR)
     COUNT = COUNT + 1
     BAR_STIFF_MAR(COUNT_MAR,i)=BAR_STIFF_MAR(COUNT_MAR,i)+
BREADTH*Db_MAR(COUNT_MAR,COUNT)*(Material_Top(COUNT_MAR,j)-
Material_Bot(COUNT_MAR,j))
     BOND_STIFF_MAR(COUNT_MAR,i)=BOND_STIFF_MAR(COUNT_MAR,i)+
D_bond_MAR(COUNT_MAR,COUNT)*PERIMETER(COUNT_MAR)*LENGTH*w(WL)
    END DO
   END DO
  END DO

END SUBROUTINE ELE_STIFF_MARNON
```

```
SUBROUTINE Global_Stiff_Matrix_MARNON
!-------------------------------------------------------------------------------
! The subroutine Global_Stiff_Matrix_MARNON
! generates global stiffness matrix.
!-------------------------------------------------------------------------------
 USE MODULE_NAME
 IMPLICIT NONE
 INTEGER :: COUNT_DIMEN, COUNT_i, COUNT_j, locate, i, j, locate_bar

 locate = 0
 locate_bar = 0
 K_Global = 0
 DO COUNT_DIMEN = 1, NUM_ELEMENT_STRUC
  CALL STIFF_MATRIX_MARNON(COUNT_DIMEN)

  DO COUNT_i = 1, DOF_ELE
   i = COUNT_i + locate
   IF (COUNT_i > 4) THEN
    i = COUNT_i + (NUM_ELEMENT_STRUC-1)*2 + locate_bar
   END IF

   IF (COUNT_i > 6) THEN
    i = i + (1+INT((COUNT_i-7)/2))*(NUM_ELEMENT_STRUC-1)
   END IF

   DO COUNT_j = 1, DOF_ELE
    j = COUNT_j + locate
    IF (COUNT_j > 4) THEN
     j = COUNT_j + (NUM_ELEMENT_STRUC-1)*2 + locate_bar
    END IF

    IF (COUNT_j > 6) THEN
     j = j + (1+INT((COUNT_j-7)/2))*(NUM_ELEMENT_STRUC-1)
    END IF

    K_Global(i,j) = K_Global(i,j) + K0(COUNT_i,COUNT_j)

   END DO
  END DO
  locate = locate + 2
  locate_bar = locate_bar + 1
 END DO

END SUBROUTINE Global_Stiff_Matrix_MARNON
```

```
SUBROUTINE Global_Stiff_Matrix_GeoNon
!---------------------------------------------------------------------------
! The subroutine Global_Stiff_Matrix_GeoNon
! generates global stiffness matrix.
!---------------------------------------------------------------------------
 USE MODULE_NAME
 IMPLICIT NONE
 INTEGER :: i, j
 INTEGER :: COUNT_DIMEN, COUNT_i, COUNT_j, locate, locate_bar

 ALLOCATE(K_Large_Global(DIMENSION_GLOBAL,DIMENSION_GLOBAL))
 ALLOCATE(K_Sigma_Global(DIMENSION_GLOBAL,DIMENSION_GLOBAL))

 locate = 0
 locate_bar = 0
 K_Large_Global = 0
 K_Sigma_Global = 0
 DO COUNT_DIMEN = 1, NUM_ELEMENT_STRUC
  CALL STIFF_MATRIX_GEONON(COUNT_DIMEN)
  CALL INITIAL_STRESS_MATRIX(COUNT_DIMEN)
  DO COUNT_i = 1, DOF_ELE
   i = COUNT_i + locate
   IF (COUNT_i > 4) THEN
    i = COUNT_i + (NUM_ELEMENT_STRUC-1)*2 + locate_bar
   END IF

   IF (COUNT_i > 6) THEN
    i = i + (1+INT((COUNT_i-7)/2))*(NUM_ELEMENT_STRUC-1)
   END IF

   DO COUNT_j = 1, DOF_ELE
    j = COUNT_j + locate
    IF (COUNT_j > 4) THEN
     j = COUNT_j + (NUM_ELEMENT_STRUC-1)*2 + locate_bar
    END IF

    IF (COUNT_j > 6) THEN
     j = j + (1+INT((COUNT_j-7)/2))*(NUM_ELEMENT_STRUC-1)
    END IF

    K_Large_Global(i,j)=K_Large_Global(i,j)+K_Large(COUNT_i,
COUNT_j)
    K_Sigma_Global(i,j)=K_Sigma_Global(i,j)+K_Sigma(COUNT_i,
COUNT_j)
   END DO
  END DO
  locate = locate + 2
  locate_bar = locate_bar + 1
 END DO

END SUBROUTINE Global_Stiff_Matrix_GeoNon
```

```
SUBROUTINE STIFF_MATRIX_MARNON(ELEMENT)
!--------------------------------------------------------------------------------
! The subroutine STIFF_MATRIX_MARNON generates
! element stiffness matrix.
!--------------------------------------------------------------------------------
 USE MODULE_NAME
 IMPLICIT NONE
 REAL, ALLOCATABLE, DIMENSION(:) :: u, w
 INTEGER :: COUNT, ELEMENT
 INTEGER :: i, j, COUNT_Gauss, COUNT_BAR
 INTEGER :: SHAPE_BAR1, SHAPE_BAR2

 {The element stiffness matrix for concrete beam has been discussed
in Appendix G. Omitted herein}

 ! FOR REINFORCEMENT
 K_BAR = 0
 K_BOND = 0
 DO COUNT_BAR = 1, NUM_BAR_ELE
  SHAPE_BAR = 0

  SHAPE_BAR1 = COUNT_BAR*2 + 3
  SHAPE_BAR2 = COUNT_BAR*2 + 4

  SHAPE_BAR(1,SHAPE_BAR1) = BAR_1
  SHAPE_BAR(1,SHAPE_BAR2) = BAR_2

  DO i = 1, DOF_ELE
   DO j = 1, DOF_ELE
    COUNT = NUM_GAUSS_POINT*(ELEMENT-1)
    DO COUNT_Gauss = 1, NUM_GAUSS_POINT
     COUNT = COUNT + 1
     K_BAR1 = MATMUL(TRANSPOSE(SHAPE_BAR),SHAPE_BAR)* BAR_STIFF_MAR
(COUNT_BAR+1,COUNT)
     K_BAR(i,j) = K_BAR(i,j) + 0.5*L*w(COUNT_Gauss)*K_BAR1(i,j)
     K_BOND1=MATMUL(TRANSPOSE(SHAPE_BAR),SHAPE_BAR)*
BOND_STIFF_MAR(COUNT_BAR+1,COUNT)
     K_BOND(i,j) = K_BOND(i,j) + 0.5*L*w(COUNT_Gauss)*K_BOND1(i,j)
    END DO
   END DO
  END DO
 END DO

 K0 = Kb + Ks + K_BAR + K_BOND

END SUBROUTINE STIFF_MATRIX_MARNON
```

Appendix I: Finite element code for composite beam element: Nonlinear analysis with temperature effect

This appendix provides the principal finite element code for nonlinear analysis of reinforced concrete beams at elevated temperatures. Only subroutines related to temperature effect are presented in this appendix. The main program and the subroutines for the analysis under ambient temperature are omitted herein.

```
PROGRAM TEMPERATURE_EFFECT
!-------------------------------------------------------------------------------
! This program is developed for nonlinear analysis
! of reinforced concrete beams at elevated
! temperatures.
!-------------------------------------------------------------------------------
 USE MODULE_NAME
 IMPLICIT NONE
 INTEGER :: STIFF_SIZE, COUNT_NUM, COUNT_j, n, m, MAX, i, j, k

 {The codes for the analysis under ambient temperature have been
previously introduced. Omitted herein}

 CALL TEM_INPUT_DATA
!-------------------------------------------------------------------------------
! The subroutine TEM_INPUT_DATA reads in the
! basic information of the model related to the
! temperature effect, including size of quadrilateral
! element, ambient temperature, internal heat
! source, convection coefficient, prescribed heat flux,
! reinforcing bar type and concrete cover thickness.
! Calculate thermal conductivity, specific heat and
! mass density at ambient temperature.
!
! This subroutine is omitted herein.
!-------------------------------------------------------------------------------

 CALL TOTAL_FORCE

 ALLOCATE(LEFT(STIFF_DIMENSION,STIFF_DIMENSION))
 ALLOCATE(RIGHT(STIFF_DIMENSION,STIFF_DIMENSION))
```

```
ALLOCATE(RIGHT_VEC(STIFF_DIMENSION,1))

STRAIN_TH_AVERAGE = 0
DO STEP_NON = 1, NUM_TIME_STEP
 Displacement_Initial = Displacement_Vec
 CALL GLOBAL_STIFFNESS_MATRIX_TEM
 CALL MASS_MATRIX
 CALL TRANSIENT_TEMPERATURE
 CALL TEMPERATURE_MATERIAL

 INITIAL_TEM = NODAL_TEM

 CALL ITERATION_NEWTON

 NONDIS_Global_Step = DIS_Global
 NONROTA_Global_Step = ROTA_Global

 DEALLOCATE(k_AVERAGE,E_AVERAGE)
 DEALLOCATE(k_ap_AVERAGE,k_au_AVERAGE)
 DEALLOCATE(STRAIN_cu_AVERAGE,STRAIN_ce_AVERAGE)
 DEALLOCATE(Db_TEM,Ds_TEM)
 DEALLOCATE(Crack_Stress_TEM,Ulti_Stress_TEM)
 DEALLOCATE(STRESS_ap,STRESS_au)
 DEALLOCATE(Crack_Strain_TEM,Tension_Strain_TEM)
END DO

END PROGRAM TEMPERATURE_EFFECT


SUBROUTINE TOTAL_FORCE
!--------------------------------------------------------------------------------
! The subroutine TOTAL_FORCE generates total
! thermal load vector
!--------------------------------------------------------------------------------
 USE MODULE_NAME
 IMPLICIT NONE
 INTEGER :: COUNT_ELEMENT, locate, COUNT_i, COUNT_LAYER

 ALLOCATE(F_TOTAL(STIFF_DIMENSION,1))

 locate = 0
 F_TOTAL = 0
 DO COUNT_ELEMENT = 1, NUM_ELEMENT_HEAT
  CALL FORCE_MATRIX(COUNT_ELEMENT)

  DO COUNT_LAYER = 1, NUM_LAYER-1
   IF(COUNT_ELEMENT==(COUNT_LAYER*NUM_ELEMENT_HEAT_LAYER+1))THEN
    locate = locate + 1
   END IF
  END DO
```

```
  DO COUNT_i = 1, 4
   IF (COUNT_i > 2) THEN
    locate = locate + NUM_ELEMENT_HEAT_LAYER - 1
   END IF

   F_TOTAL(COUNT_i+locate,1) = F_TOTAL(COUNT_i+locate,1) + f_SOURCE
(COUNT_i,1) + f_FLUX(COUNT_i,1) + f_CONVECTION(COUNT_i,1)

   IF (COUNT_i > 2) THEN
    locate = locate - NUM_ELEMENT_HEAT_LAYER + 1
   END IF
   END DO

  locate = locate + 1

 END DO

END SUBROUTINE TOTAL_FORCE


SUBROUTINE FORCE_MATRIX(ELEMENT)
!--------------------------------------------------------------------------------
! The subroutine FORCE_MATRIX generates
! components of thermal load vector
!--------------------------------------------------------------------------------
 USE MODULE_NAME
 IMPLICIT NONE
 REAL :: s1, s2, t1, t2
 INTEGER :: i, j, COUNT_GAUSS, COUNT_GAUSS1, COUNT_GAUSS2, ELEMENT,
COUNT_DIFF
 REAL, ALLOCATABLE, DIMENSION(:) :: u, w
 INTEGER :: TWO_SIDE, BL, BR, TL, TR

 CALL Gauss_Beam(u,w)

 f_SOURCE = 0
 DO i = 1,4
  DO COUNT_GAUSS1 = 1, Num_Gauss_Point_TEM
   DO COUNT_GAUSS2 = 1, Num_Gauss_Point_TEM
    s = 0.5*(s2-s1)*u(COUNT_GAUSS2) + 0.5*(s2+s1)
    t = 0.5*(t2-t1)*u(COUNT_GAUSS1) + 0.5*(t2+t1)
    CALL SHAPE_FUNCTION_TEM
    f_SOURCE1 = HEAT_SOURCE_ELE(ELEMENT)*THICKNESS_TEM*TRANSPOSE(N)
    f_SOURCE(i,1) = f_SOURCE(i,1) + 0.5*(t2-t1)*0.5*(s2-s1)*
w(COUNT_GAUSS2)*w(COUNT_GAUSS1)*f_SOURCE1(i,1)
   END DO
  END DO
 END DO

 DO COUNT_DIFF = 1, NUM_DIFF
  IF (ELEMENT == DIFF_ELE(COUNT_DIFF)) THEN
```

```
   ! BOTTOM_LEFT ELEMENT
   IF (ELEMENT == 1) THEN
    WRITE(*,*) 'Does BOTTOM_LEFT ELEMENT convect from two sides?
(Yes == 1; No == 0)'
    READ(*,*) TWO_SIDE

    IF (TWO_SIDE == 1) THEN
     s = -0.5*BREADTH_HEAT

     f_FLUX = 0
     f_CONVECTION = 0
     DO i = 1,4
      DO COUNT_GAUSS = 1, Num_Gauss_Point_TEM
       t = 0.5*(t2-t1)*u(COUNT_GAUSS) + 0.5*(t2+t1)
       CALL SHAPE_FUNCTION_TEM
       f_FLUX1 = HEAT_FLUX_ELE(ELEMENT)*THICKNESS_TEM*TRANSPOSE(N)
       f_FLUX(i,1)=f_FLUX(i,1)+0.5*(t2-t1)*w(COUNT_GAUSS)*
f_FLUX1(i,1)
       f_CONVECTION1 = CONVECTION_COEFFI_ELE(ELEMENT)*FREE_TEM_ELE
(ELEMENT)*THICKNESS_TEM*TRANSPOSE(N)
       f_CONVECTION(i,1)=f_CONVECTION(i,1)+0.5*(t2-t1)*
w(COUNT_GAUSS)*f_CONVECTION1(i,1)
      END DO
     END DO

     t = -0.5*HEIGHT_HEAT

     DO i = 1,4
      DO COUNT_GAUSS = 1, Num_Gauss_Point_TEM
       s = 0.5*(s2-s1)*u(COUNT_GAUSS) + 0.5*(s2+s1)
       CALL SHAPE_FUNCTION_TEM
       f_FLUX1 = HEAT_FLUX_ELE(ELEMENT)*THICKNESS_TEM*TRANSPOSE(N)
       f_FLUX(i,1)=f_FLUX(i,1)+0.5*(s2-s1)*w(COUNT_GAUSS)*
f_FLUX1(i,1)
       f_CONVECTION1 = CONVECTION_COEFFI_ELE(ELEMENT)*FREE_TEM_ELE
(ELEMENT)*THICKNESS_TEM*TRANSPOSE(N)
       f_CONVECTION(i,1)=f_CONVECTION(i,1)+0.5*(s2-s1)*
w(COUNT_GAUSS)*f_CONVECTION1(i,1)
      END DO
     END DO
    ELSE
     WRITE(*,*) 'Does the element convect from bottom side or left
side? (BOTTOM = 1; LEFT = 0)'
     READ(*,*) BL
    END IF
   END IF

   ! BOTTOM_RIGHT ELEMENT
   IF (ELEMENT == NUM_ELEMENT_HEAT_LAYER) THEN
    WRITE(*,*) 'Does BOTTOM_RIGHT ELEMENT convect from two sides?
(Yes == 1; No == 0)'
```

```
    READ(*,*) TWO_SIDE

    IF (TWO_SIDE == 1) THEN
     s = 0.5*BREADTH_HEAT

     f_FLUX = 0
     f_CONVECTION = 0
     DO i = 1,4
      DO COUNT_GAUSS = 1, Num_Gauss_Point_TEM
       t = 0.5*(t2-t1)*u(COUNT_GAUSS) + 0.5*(t2+t1)
       CALL SHAPE_FUNCTION_TEM
       f_FLUX1 = HEAT_FLUX_ELE(ELEMENT)*THICKNESS_TEM*TRANSPOSE(N)
       f_FLUX(i,1)=f_FLUX(i,1)+0.5*(t2-t1)*w(COUNT_GAUSS)*
f_FLUX1(i,1)
       f_CONVECTION1 = CONVECTION_COEFFI_ELE(ELEMENT)*FREE_TEM_ELE
(ELEMENT)*THICKNESS_TEM*TRANSPOSE(N)
       f_CONVECTION(i,1)=f_CONVECTION(i,1)+0.5*(t2-t1)*
w(COUNT_GAUSS)*f_CONVECTION1(i,1)
      END DO
     END DO

     t = -0.5*HEIGHT_HEAT

     DO i = 1,4
      DO COUNT_GAUSS = 1, Num_Gauss_Point_TEM
       s = 0.5*(s2-s1)*u(COUNT_GAUSS) + 0.5*(s2+s1)
       CALL SHAPE_FUNCTION_TEM
       f_FLUX1 = HEAT_FLUX_ELE(ELEMENT)*THICKNESS_TEM*TRANSPOSE(N)
       f_FLUX(i,1)=f_FLUX(i,1)+0.5*(s2-s1)*w(COUNT_GAUSS)*
f_FLUX1(i,1)
       f_CONVECTION1 = CONVECTION_COEFFI_ELE(ELEMENT)*FREE_TEM_ELE
(ELEMENT)*THICKNESS_TEM*TRANSPOSE(N)
       f_CONVECTION(i,1)=f_CONVECTION(i,1)+0.5*(s2-s1)*
w(COUNT_GAUSS)*f_CONVECTION1(i,1)
      END DO
     END DO
    ELSE
     WRITE(*,*) 'Does the element convect from bottom side or right
side? (BOTTOM = 1; RIGHT = 0)'
     READ(*,*) BR
    END IF
   END IF

   ! TOP_LEFT ELEMENT
   IF (ELEMENT == NUM_ELEMENT_HEAT_LAYER*(NUM_LAYER-1)+1) THEN
    WRITE(*,*) 'Does TOP_LEFT ELEMENT convect from two sides?
(Yes == 1; No == 0)'
    READ(*,*) TWO_SIDE

    IF (TWO_SIDE == 1) THEN
     s = -0.5*BREADTH_HEAT
```

```
     f_FLUX = 0
     f_CONVECTION = 0
     DO i = 1,4
      DO COUNT_GAUSS = 1, Num_Gauss_Point_TEM
       t = 0.5*(t2-t1)*u(COUNT_GAUSS) + 0.5*(t2+t1)
       CALL SHAPE_FUNCTION_TEM
       f_FLUX1 = HEAT_FLUX_ELE(ELEMENT)*THICKNESS_TEM*TRANSPOSE(N)
       f_FLUX(i,1)=f_FLUX(i,1)+0.5*(t2-t1)*w(COUNT_GAUSS)*
f_FLUX1(i,1)
       f_CONVECTION1=CONVECTION_COEFFI_ELE(ELEMENT)*FREE_TEM_ELE
(ELEMENT)*THICKNESS_TEM*TRANSPOSE(N)
       f_CONVECTION(i,1)=f_CONVECTION(i,1)+0.5*(t2-t1)*
w(COUNT_GAUSS)*f_CONVECTION1(i,1)
      END DO
     END DO

     t = 0.5*HEIGHT_HEAT

     DO i = 1,4
      DO COUNT_GAUSS = 1, Num_Gauss_Point_TEM
       s = 0.5*(s2-s1)*u(COUNT_GAUSS) + 0.5*(s2+s1)
       CALL SHAPE_FUNCTION_TEM
       f_FLUX1 = HEAT_FLUX_ELE(ELEMENT)*THICKNESS_TEM*TRANSPOSE(N)
       f_FLUX(i,1)=f_FLUX(i,1)+0.5*(s2-s1)*w(COUNT_GAUSS)*
f_FLUX1(i,1)
       f_CONVECTION1=CONVECTION_COEFFI_ELE(ELEMENT)*FREE_TEM_ELE
(ELEMENT)*THICKNESS_TEM*TRANSPOSE(N)
       f_CONVECTION(i,1)=f_CONVECTION(i,1)+0.5*(s2-s1)*
w(COUNT_GAUSS)*f_CONVECTION1(i,1)
      END DO
     END DO
    ELSE
     WRITE(*,*) 'Does the element convect from top side or left
side? (TOP = 1; LEFT = 0)'
     READ(*,*) TL
    END IF
   END IF

   ! TOP_RIGHT ELEMENT
   IF (ELEMENT == NUM_ELEMENT_HEAT) THEN
    WRITE(*,*) 'Does TOP_RIGHT ELEMENT convect from two sides?
(Yes == 1; No == 0)'
    READ(*,*) TWO_SIDE

    IF (TWO_SIDE == 1) THEN
     s = 0.5*BREADTH_HEAT

     f_FLUX = 0
     f_CONVECTION = 0
     DO i = 1,4
      DO COUNT_GAUSS = 1, Num_Gauss_Point_TEM
```

```
      t = 0.5*(t2-t1)*u(COUNT_GAUSS) + 0.5*(t2+t1)
      CALL SHAPE_FUNCTION_TEM
      f_FLUX1 = HEAT_FLUX_ELE(ELEMENT)*THICKNESS_TEM*TRANSPOSE(N)
      f_FLUX(i,1)=f_FLUX(i,1)+0.5*(t2-t1)*w(COUNT_GAUSS)*
f_FLUX1(i,1)
      f_CONVECTION1=CONVECTION_COEFFI_ELE(ELEMENT)*FREE_TEM_ELE
(ELEMENT)*THICKNESS_TEM*TRANSPOSE(N)
      f_CONVECTION(i,1)=f_CONVECTION(i,1)+0.5*(t2-t1)*
w(COUNT_GAUSS)*f_CONVECTION1(i,1)
     END DO
    END DO

    t = 0.5*HEIGHT_HEAT

    DO i = 1,4
     DO COUNT_GAUSS = 1, Num_Gauss_Point_TEM
      s = 0.5*(s2-s1)*u(COUNT_GAUSS) + 0.5*(s2+s1)
      CALL SHAPE_FUNCTION_TEM
      f_FLUX1 = HEAT_FLUX_ELE(ELEMENT)*THICKNESS_TEM*TRANSPOSE(N)
      f_FLUX(i,1)=f_FLUX(i,1)+0.5*(s2-s1)*w(COUNT_GAUSS)*
f_FLUX1(i,1)
      f_CONVECTION1 = CONVECTION_COEFFI_ELE(ELEMENT)*FREE_TEM_ELE
(ELEMENT)*THICKNESS_TEM*TRANSPOSE(N)
      f_CONVECTION(i,1)=f_CONVECTION(i,1)+0.5*(s2-s1)*
w(COUNT_GAUSS)*f_CONVECTION1(i,1)
     END DO
    END DO
   ELSE
    WRITE(*,*) 'Does the element convect from top side or right
side? (TOP = 1; RIGHT = 0)'
    READ(*,*) TR
   END IF
  END IF

  ! TOP ELEMENTS
  IF (INT((ELEMENT-1)/NUM_ELEMENT_HEAT_LAYER) >= (NUM_LAYER-1))
THEN
   IF (ELEMENT == NUM_ELEMENT_HEAT_LAYER*(NUM_LAYER-1)+1) THEN
    IF (TL == 1) THEN
     t = 0.5*HEIGHT_HEAT

     f_FLUX = 0
     f_CONVECTION = 0
     DO i = 1,4
      DO COUNT_GAUSS = 1, Num_Gauss_Point_TEM
       s = 0.5*(s2-s1)*u(COUNT_GAUSS) + 0.5*(s2+s1)
       CALL SHAPE_FUNCTION_TEM
       f_FLUX1 = HEAT_FLUX_ELE(ELEMENT)*THICKNESS_TEM*TRANSPOSE(N)
       f_FLUX(i,1)=f_FLUX(i,1)+0.5*(s2-s1)*w(COUNT_GAUSS)*
f_FLUX1(i,1)
```

```
        f_CONVECTION1 = CONVECTION_COEFFI_ELE(ELEMENT)*FREE_TEM_ELE
(ELEMENT)*THICKNESS_TEM*TRANSPOSE(N)
        f_CONVECTION(i,1)=f_CONVECTION(i,1)+0.5*(s2-s1)*
w(COUNT_GAUSS)*f_CONVECTION1(i,1)
       END DO
      END DO
     END IF
    END IF

    IF (ELEMENT == NUM_ELEMENT_HEAT) THEN
     IF (TR == 1) THEN
      t = 0.5*HEIGHT_HEAT

      f_FLUX = 0
      f_CONVECTION = 0
      DO i = 1,4
       DO COUNT_GAUSS = 1, Num_Gauss_Point_TEM
        s = 0.5*(s2-s1)*u(COUNT_GAUSS) + 0.5*(s2+s1)
        CALL SHAPE_FUNCTION_TEM
        f_FLUX1 = HEAT_FLUX_ELE(ELEMENT)*THICKNESS_TEM*TRANSPOSE(N)
        f_FLUX(i,1)=f_FLUX(i,1)+0.5*(s2-s1)*w(COUNT_GAUSS)*
f_FLUX1(i,1)
        f_CONVECTION1 = CONVECTION_COEFFI_ELE(ELEMENT)*FREE_TEM_ELE
(ELEMENT)*THICKNESS_TEM*TRANSPOSE(N)
        f_CONVECTION(i,1)=f_CONVECTION(i,1)+0.5*(s2-s1)*
w(COUNT_GAUSS)*f_CONVECTION1(i,1)
       END DO
      END DO
     END IF
    END IF

    IF (ELEMENT /= NUM_ELEMENT_HEAT_LAYER*(NUM_LAYER-1)+1) THEN
     IF (ELEMENT /= NUM_ELEMENT_HEAT) THEN

      t = 0.5*HEIGHT_HEAT

      f_FLUX = 0
      f_CONVECTION = 0
      DO i = 1,4
       DO COUNT_GAUSS = 1, Num_Gauss_Point_TEM
        s = 0.5*(s2-s1)*u(COUNT_GAUSS) + 0.5*(s2+s1)
        CALL SHAPE_FUNCTION_TEM
        f_FLUX1 = HEAT_FLUX_ELE(ELEMENT)*THICKNESS_TEM*TRANSPOSE(N)
        f_FLUX(i,1)=f_FLUX(i,1)+0.5*(s2-s1)*w(COUNT_GAUSS)*
f_FLUX1(i,1)
        f_CONVECTION1 = CONVECTION_COEFFI_ELE(ELEMENT)*FREE_TEM_ELE
(ELEMENT)*THICKNESS_TEM*TRANSPOSE(N)
        f_CONVECTION(i,1)=f_CONVECTION(i,1)+0.5*(s2-s1)*
w(COUNT_GAUSS)*f_CONVECTION1(i,1)
       END DO
      END DO
```

```
     END IF
    END IF
   END IF

   ! BOTTOM ELEMENTS
   IF (INT((ELEMENT-1)/NUM_ELEMENT_HEAT_LAYER) <= 0) THEN
    IF (ELEMENT == 1) THEN
     IF (BL == 1) THEN
      t = -0.5*HEIGHT_HEAT

      f_FLUX = 0
      f_CONVECTION = 0
      DO i = 1,4
       DO COUNT_GAUSS = 1, Num_Gauss_Point_TEM
        s = 0.5*(s2-s1)*u(COUNT_GAUSS) + 0.5*(s2+s1)
        CALL SHAPE_FUNCTION_TEM
        f_FLUX1 = HEAT_FLUX_ELE(ELEMENT)*THICKNESS_TEM*TRANSPOSE(N)
        f_FLUX(i,1)=f_FLUX(i,1)+0.5*(s2-s1)*w(COUNT_GAUSS)*
f_FLUX1(i,1)
        f_CONVECTION1 = CONVECTION_COEFFI_ELE(ELEMENT)*FREE_TEM_ELE
(ELEMENT)*THICKNESS_TEM*TRANSPOSE(N)
        f_CONVECTION(i,1) = f_CONVECTION(i,1) + 0.5*(s2-s1)*
w(COUNT_GAUSS)*f_CONVECTION1(i,1)
       END DO
      END DO
     END IF
    END IF

    IF (ELEMENT == NUM_ELEMENT_HEAT_LAYER) THEN
     IF (BR == 1) THEN
      t = -0.5*HEIGHT_HEAT

      f_FLUX = 0
      f_CONVECTION = 0
      DO i = 1,4
       DO COUNT_GAUSS = 1, Num_Gauss_Point_TEM
        s = 0.5*(s2-s1)*u(COUNT_GAUSS) + 0.5*(s2+s1)
        CALL SHAPE_FUNCTION_TEM
        f_FLUX1 = HEAT_FLUX_ELE(ELEMENT)*THICKNESS_TEM*TRANSPOSE(N)
        f_FLUX(i,1) = f_FLUX(i,1) + 0.5*(s2-s1)*w(COUNT_GAUSS)*
f_FLUX1(i,1)
        f_CONVECTION1 = CONVECTION_COEFFI_ELE(ELEMENT)*FREE_TEM_ELE
(ELEMENT)*THICKNESS_TEM*TRANSPOSE(N)
        f_CONVECTION(i,1) = f_CONVECTION(i,1) + 0.5*(s2-s1)*
w(COUNT_GAUSS)*f_CONVECTION1(i,1)
       END DO
      END DO
     END IF
    END IF
```

```
    IF (ELEMENT /= 1) THEN
     IF (ELEMENT /= NUM_ELEMENT_HEAT_LAYER) THEN
      t = -0.5*HEIGHT_HEAT

      f_FLUX = 0
      f_CONVECTION = 0
      DO i = 1,4
       DO COUNT_GAUSS = 1, Num_Gauss_Point_TEM
        s = 0.5*(s2-s1)*u(COUNT_GAUSS) + 0.5*(s2+s1)
        CALL SHAPE_FUNCTION_TEM
        f_FLUX1 = HEAT_FLUX_ELE(ELEMENT)*THICKNESS_TEM*TRANSPOSE(N)
        f_FLUX(i,1) = f_FLUX(i,1) + 0.5*(s2-s1)*w(COUNT_GAUSS)*
f_FLUX1(i,1)
        f_CONVECTION1 = CONVECTION_COEFFI_ELE(ELEMENT)*FREE_TEM_ELE
(ELEMENT)*THICKNESS_TEM*TRANSPOSE(N)
        f_CONVECTION(i,1) = f_CONVECTION(i,1) + 0.5*(s2-s1)*
w(COUNT_GAUSS)*f_CONVECTION1(i,1)
       END DO
      END DO
     END IF
    END IF
   END IF

   ! LEFT ELEMENTS
   LEFT_ELEMENT = 1
   DO WHILE (LEFT_ELEMENT <= NUM_ELEMENT_HEAT)
    IF (ELEMENT == LEFT_ELEMENT) THEN
     IF (ELEMENT == 1) THEN
      IF (BL == 0) THEN
       s = -0.5*BREADTH_HEAT

       f_FLUX = 0
       f_CONVECTION = 0
       DO i = 1,4
        DO COUNT_GAUSS = 1, Num_Gauss_Point_TEM
         t = 0.5*(t2-t1)*u(COUNT_GAUSS) + 0.5*(t2+t1)
         CALL SHAPE_FUNCTION_TEM
         f_FLUX1 = HEAT_FLUX_ELE(ELEMENT)*THICKNESS_TEM*TRANSPOSE(N)
         f_FLUX(i,1) = f_FLUX(i,1) + 0.5*(t2-t1)*w(COUNT_GAUSS)*
f_FLUX1(i,1)
         f_CONVECTION1 = CONVECTION_COEFFI_ELE(ELEMENT)*
FREE_TEM_ELE(ELEMENT)*THICKNESS_TEM*TRANSPOSE(N)
         f_CONVECTION(i,1) = f_CONVECTION(i,1) + 0.5*(t2-t1)*
w(COUNT_GAUSS)*f_CONVECTION1(i,1)
        END DO
       END DO
      END IF
     END IF
```

```
      IF (ELEMENT == NUM_ELEMENT_HEAT_LAYER*(NUM_LAYER-1)+1) THEN
       IF (TL == 0) THEN
        s = -0.5*BREADTH_HEAT

        f_FLUX = 0
        f_CONVECTION = 0
        DO i = 1,4
         DO COUNT_GAUSS = 1, Num_Gauss_Point_TEM
          t = 0.5*(t2-t1)*u(COUNT_GAUSS) + 0.5*(t2+t1)
          CALL SHAPE_FUNCTION_TEM
          f_FLUX1=HEAT_FLUX_ELE(ELEMENT)*THICKNESS_TEM*TRANSPOSE(N)
          f_FLUX(i,1) = f_FLUX(i,1) + 0.5*(t2-t1)*w(COUNT_GAUSS)*
f_FLUX1(i,1)
          f_CONVECTION1 = CONVECTION_COEFFI_ELE(ELEMENT)*
FREE_TEM_ELE(ELEMENT)*THICKNESS_TEM*TRANSPOSE(N)
          f_CONVECTION(i,1) = f_CONVECTION(i,1) + 0.5*(t2-t1)*
w(COUNT_GAUSS)*f_CONVECTION1(i,1)
         END DO
        END DO
       END IF
      END IF

      IF (ELEMENT /= 1) THEN
       IF (ELEMENT /= NUM_ELEMENT_HEAT_LAYER*(NUM_LAYER-1)+1) THEN
        s = -0.5*BREADTH_HEAT

        f_FLUX = 0
        f_CONVECTION = 0
        DO i = 1,4
         DO COUNT_GAUSS = 1, Num_Gauss_Point_TEM
          t = 0.5*(t2-t1)*u(COUNT_GAUSS) + 0.5*(t2+t1)
          CALL SHAPE_FUNCTION_TEM
          f_FLUX1=HEAT_FLUX_ELE(ELEMENT)*THICKNESS_TEM*TRANSPOSE(N)
          f_FLUX(i,1) = f_FLUX(i,1) + 0.5*(t2-t1)*w(COUNT_GAUSS)*
f_FLUX1(i,1)
          f_CONVECTION1 = CONVECTION_COEFFI_ELE(ELEMENT)*
FREE_TEM_ELE(ELEMENT)*THICKNESS_TEM*TRANSPOSE(N)
          f_CONVECTION(i,1) = f_CONVECTION(i,1) + 0.5*(t2-t1)*
w(COUNT_GAUSS)*f_CONVECTION1(i,1)
         END DO
        END DO
       END IF
      END IF
     END IF

     LEFT_ELEMENT = LEFT_ELEMENT + NUM_ELEMENT_HEAT_LAYER

    END DO

    ! RIGHT ELEMENTS
    RIGHT_ELEMENT = NUM_ELEMENT_HEAT_LAYER
```

```
   DO WHILE (RIGHT_ELEMENT <= NUM_ELEMENT_HEAT)
    IF (ELEMENT == RIGHT_ELEMENT) THEN
     IF (ELEMENT == NUM_ELEMENT_HEAT_LAYER) THEN
      IF (BR == 0) THEN
       s = 0.5*BREADTH_HEAT

       f_FLUX = 0
       f_CONVECTION = 0
       DO i = 1,4
        DO COUNT_GAUSS = 1, Num_Gauss_Point_TEM
         t = 0.5*(t2-t1)*u(COUNT_GAUSS) + 0.5*(t2+t1)
         CALL SHAPE_FUNCTION_TEM
         f_FLUX1=HEAT_FLUX_ELE(ELEMENT)*THICKNESS_TEM*TRANSPOSE(N)
         f_FLUX(i,1) = f_FLUX(i,1) + 0.5*(t2-t1)*w(COUNT_GAUSS)*
f_FLUX1(i,1)
         f_CONVECTION1 = CONVECTION_COEFFI_ELE(ELEMENT)*
FREE_TEM_ELE(ELEMENT)*THICKNESS_TEM*TRANSPOSE(N)
         f_CONVECTION(i,1) = f_CONVECTION(i,1) + 0.5*(t2-t1)*
w(COUNT_GAUSS)*f_CONVECTION1(i,1)
        END DO
       END DO
      END IF
     END IF

     IF (ELEMENT == NUM_ELEMENT_HEAT) THEN
      IF (TR == 0) THEN
       s = 0.5*BREADTH_HEAT

       f_FLUX = 0
       f_CONVECTION = 0
       DO i = 1,4
        DO COUNT_GAUSS = 1, Num_Gauss_Point_TEM
         t = 0.5*(t2-t1)*u(COUNT_GAUSS) + 0.5*(t2+t1)
         CALL SHAPE_FUNCTION_TEM
         f_FLUX1=HEAT_FLUX_ELE(ELEMENT)*THICKNESS_TEM*TRANSPOSE(N)
         f_FLUX(i,1) = f_FLUX(i,1) + 0.5*(t2-t1)*w(COUNT_GAUSS)*
f_FLUX1(i,1)
         f_CONVECTION1 = CONVECTION_COEFFI_ELE(ELEMENT)*
FREE_TEM_ELE(ELEMENT)*THICKNESS_TEM*TRANSPOSE(N)
         f_CONVECTION(i,1) = f_CONVECTION(i,1) + 0.5*(t2-t1)*
w(COUNT_GAUSS)*f_CONVECTION1(i,1)
        END DO
       END DO
      END IF
     END IF

     IF (ELEMENT /= NUM_ELEMENT_HEAT_LAYER) THEN
      IF (ELEMENT /= NUM_ELEMENT_HEAT) THEN
       s = 0.5*BREADTH_HEAT
```

```
       f_FLUX = 0
       f_CONVECTION = 0
       DO i = 1,4
        DO COUNT_GAUSS = 1, Num_Gauss_Point_TEM
         t = 0.5*(t2-t1)*u(COUNT_GAUSS) + 0.5*(t2+t1)
         CALL SHAPE_FUNCTION_TEM
         f_FLUX1=HEAT_FLUX_ELE(ELEMENT)*THICKNESS_TEM*TRANSPOSE(N)
         f_FLUX(i,1) = f_FLUX(i,1) + 0.5*(t2-t1)*w(COUNT_GAUSS)*
f_FLUX1(i,1)
         f_CONVECTION1 = CONVECTION_COEFFI_ELE(ELEMENT)*
FREE_TEM_ELE(ELEMENT)*THICKNESS_TEM*TRANSPOSE(N)
         f_CONVECTION(i,1) = f_CONVECTION(i,1) + 0.5*(t2-t1)*
w(COUNT_GAUSS)*f_CONVECTION1(i,1)
        END DO
       END DO
      END IF
     END IF
    END IF

    RIGHT_ELEMENT = RIGHT_ELEMENT + NUM_ELEMENT_HEAT_LAYER

   END DO
  END IF
 END DO

END SUBROUTINE FORCE_MATRIX


SUBROUTINE SHAPE_FUNCTION_TEM
!--------------------------------------------------------------------------------
! The subroutine SHAPE_FUNCTION_TEM provides
! shape function for quadrilateral element
!--------------------------------------------------------------------------------
 USE MODULE_NAME

 N(1,1)=(0.5*BREADTH_HEAT-s)*(0.5*HEIGHT_HEAT-t)/(BREADTH_HEAT*
HEIGHT_HEAT)
 N(1,2)=(0.5*BREADTH_HEAT+s)*(0.5*HEIGHT_HEAT-t)/(BREADTH_HEAT*
HEIGHT_HEAT)
 N(1,3)=(0.5*BREADTH_HEAT-s)*(0.5*HEIGHT_HEAT+t)/(BREADTH_HEAT*
HEIGHT_HEAT)
 N(1,4)=(0.5*BREADTH_HEAT+s)*(0.5*HEIGHT_HEAT+t)/(BREADTH_HEAT*
HEIGHT_HEAT)

 B(1,1) = (-0.5*HEIGHT_HEAT+t)/(BREADTH_HEAT*HEIGHT_HEAT)
 B(1,2) = (0.5*HEIGHT_HEAT-t)/(BREADTH_HEAT*HEIGHT_HEAT)
 B(1,3) = (-0.5*HEIGHT_HEAT-t)/(BREADTH_HEAT*HEIGHT_HEAT)
 B(1,4) = (0.5*HEIGHT_HEAT+t)/(BREADTH_HEAT*HEIGHT_HEAT)
 B(2,1) = (-0.5*BREADTH_HEAT+s)/(BREADTH_HEAT*HEIGHT_HEAT)
 B(2,2) = (-0.5*BREADTH_HEAT-s)/(BREADTH_HEAT*HEIGHT_HEAT)
```

```
 B(2,3) = (0.5*BREADTH_HEAT-s)/(BREADTH_HEAT*HEIGHT_HEAT)
 B(2,4) = (0.5*BREADTH_HEAT+s)/(BREADTH_HEAT*HEIGHT_HEAT)

END SUBROUTINE SHAPE_FUNCTION_TEM


SUBROUTINE GLOBAL_STIFFNESS_MATRIX_TEM
!--------------------------------------------------------------------------------
! The subroutine GLOBAL_STIFFNESS_MATRIX_TEM
! generates global conduction matrix
!--------------------------------------------------------------------------------
 USE MODULE_NAME
 IMPLICIT NONE
 INTEGER :: COUNT_ELEMENT, COUNT_LAYER
 INTEGER :: COUNT_i, COUNT_j, locate1, locate2

 locate1 = 0
 locate2 = 0
 K_TEM_GLOBAL = 0
 DO COUNT_ELEMENT = 1, NUM_ELEMENT_HEAT
  CALL ELEMENT_STIFFNESS_HEAT(COUNT_ELEMENT)

  DO COUNT_LAYER = 1, NUM_LAYER-1
   IF(COUNT_ELEMENT==(COUNT_LAYER*NUM_ELEMENT_HEAT_LAYER+1))THEN
    locate1 = locate1 + 1
    locate2 = locate2 + 1
   END IF
  END DO

  DO COUNT_i = 1, 4
   IF (COUNT_i > 2) THEN
    locate1 = locate1 + NUM_ELEMENT_HEAT_LAYER - 1
   END IF

   DO COUNT_j = 1, 4
    IF (COUNT_j > 2) THEN
     locate2 = locate2 + NUM_ELEMENT_HEAT_LAYER - 1
    END IF

    K_TEM_GLOBAL(COUNT_i+locate1,COUNT_j+locate2) = K_TEM_GLOBAL
 (COUNT_i+locate1,COUNT_j+locate2) + kc(COUNT_i,COUNT_j) +
 kh(COUNT_i,COUNT_j)

    IF (COUNT_j > 2) THEN
     locate2 = locate2 - NUM_ELEMENT_HEAT_LAYER + 1
    END IF
   END DO

   IF (COUNT_i > 2) THEN
    locate1 = locate1 - NUM_ELEMENT_HEAT_LAYER + 1
   END IF
  END DO
```

```
  locate1 = locate1 + 1
  locate2 = locate2 + 1

 END DO

END SUBROUTINE GLOBAL_STIFFNESS_MATRIX_TEM


SUBROUTINE ELEMENT_STIFFNESS_HEAT(ELEMENT)
!------------------------------------------------------------------------------
! The subroutine ELEMENT_STIFFNESS_HEAT
! generates element conduction matrix
!------------------------------------------------------------------------------
 USE MODULE_NAME
 IMPLICIT NONE
 REAL :: s1, s2, t1, t2
 INTEGER :: i, j, COUNT_GAUSS, COUNT_GAUSS1, COUNT_GAUSS2, ELEMENT,
COUNT_DIFF
 INTEGER :: BOTTOM_LEFT, BOTTOM_RIGHT, TOP_LEFT, TOP_RIGHT,
TWO_SIDE
 INTEGER :: BL, BR, TL, TR
 REAL, ALLOCATABLE, DIMENSION(:) :: u, w

 STIFF_CONDUCT(1,1) = THERMAL_CONDUCT_ELE_x(ELEMENT)
 STIFF_CONDUCT(1,2) = 0
 STIFF_CONDUCT(2,1) = 0
 STIFF_CONDUCT(2,2) = THERMAL_CONDUCT_ELE_y(ELEMENT)

 CALL Gauss_Beam(u,w)

 s1 = -0.5*BREADTH_HEAT
 s2 = 0.5*BREADTH_HEAT
 t1 = -0.5*HEIGHT_HEAT
 t2 = 0.5*HEIGHT_HEAT

 kc = 0
 DO i = 1,4
  DO j = 1,4
   DO COUNT_GAUSS1 = 1, Num_Gauss_Point_TEM
    DO COUNT_GAUSS2 = 1, Num_Gauss_Point_TEM
     s = 0.5*(s2-s1)*u(COUNT_GAUSS2) + 0.5*(s2+s1)
     t = 0.5*(t2-t1)*u(COUNT_GAUSS1) + 0.5*(t2+t1)
     CALL SHAPE_FUNCTION_TEM
     kc1=THICKNESS_TEM*MATMUL(MATMUL(TRANSPOSE(B),STIFF_CONDUCT),B)
     kc(i,j) = kc(i,j) + 0.5*(t2-t1)*0.5*(s2-s1)*w(COUNT_GAUSS2)*
w(COUNT_GAUSS1)*kc1(i,j)
    END DO
   END DO
  END DO
 END DO
```

```
kh = 0
DO COUNT_DIFF = 1, NUM_DIFF
 IF (ELEMENT == DIFF_ELE(COUNT_DIFF)) THEN

  ! BOTTOM_LEFT ELEMENT
  IF (ELEMENT == 1) THEN
   WRITE(*,*) 'Does BOTTOM_LEFT ELEMENT convect from two sides?
(Yes == 1; No == 0)'
   READ(*,*) TWO_SIDE

   IF (TWO_SIDE == 1) THEN
    s = -0.5*BREADTH_HEAT

    kh = 0
    DO i = 1,4
     DO j = 1,4
      DO COUNT_GAUSS = 1, Num_Gauss_Point_TEM
       t = 0.5*(t2-t1)*u(COUNT_GAUSS) + 0.5*(t2+t1)
       CALL SHAPE_FUNCTION_TEM
       kh1  =  CONVECTION_COEFFI_ELE(ELEMENT)*THICKNESS_TEM*MATMUL
(TRANSPOSE(N),N)
       kh(i,j) = kh(i,j) + 0.5*(t2-t1)*w(COUNT_GAUSS)*kh1(i,j)
      END DO
     END DO
    END DO

    t = -0.5*HEIGHT_HEAT

    DO i = 1,4
     DO j = 1,4
      DO COUNT_GAUSS = 1, Num_Gauss_Point_TEM
       s = 0.5*(s2-s1)*u(COUNT_GAUSS) + 0.5*(s2+s1)
       CALL SHAPE_FUNCTION_TEM
       kh1  =  CONVECTION_COEFFI_ELE(ELEMENT)*THICKNESS_TEM*MATMUL
(TRANSPOSE(N),N)
       kh(i,j) = kh(i,j) + 0.5*(s2-s1)*w(COUNT_GAUSS)*kh1(i,j)
      END DO
     END DO
    END DO
   ELSE
    WRITE(*,*) 'Does the element convect from bottom side or left
side? (BOTTOM = 1; LEFT = 0)'
    READ(*,*) BL
   END IF
  END IF

  ! BOTTOM_RIGHT ELEMENT
  IF (ELEMENT == NUM_ELEMENT_HEAT_LAYER) THEN
   WRITE(*,*) 'Does BOTTOM_RIGHT ELEMENT convect from two sides?
(Yes == 1; No == 0)'
   READ(*,*) TWO_SIDE
```

```
    IF (TWO_SIDE == 1) THEN
     s = 0.5*BREADTH_HEAT

     kh = 0
     DO i = 1,4
      DO j = 1,4
       DO COUNT_GAUSS = 1, Num_Gauss_Point_TEM
        t = 0.5*(t2-t1)*u(COUNT_GAUSS) + 0.5*(t2+t1)
        CALL SHAPE_FUNCTION_TEM
        kh1  =  CONVECTION_COEFFI_ELE(ELEMENT)*THICKNESS_TEM*MATMUL
(TRANSPOSE(N),N)
        kh(i,j) = kh(i,j) + 0.5*(t2-t1)*w(COUNT_GAUSS)*kh1(i,j)
       END DO
      END DO
     END DO

     t = -0.5*HEIGHT_HEAT

     DO i = 1,4
      DO j = 1,4
       DO COUNT_GAUSS = 1, Num_Gauss_Point_TEM
        s = 0.5*(s2-s1)*u(COUNT_GAUSS) + 0.5*(s2+s1)
        CALL SHAPE_FUNCTION_TEM
        kh1  =  CONVECTION_COEFFI_ELE(ELEMENT)*THICKNESS_TEM*MATMUL
(TRANSPOSE(N),N)
        kh(i,j) = kh(i,j) + 0.5*(s2-s1)*w(COUNT_GAUSS)*kh1(i,j)
       END DO
      END DO
     END DO
    ELSE
     WRITE(*,*) 'Does the element convect from bottom side or right
side? (BOTTOM = 1; RIGHT = 0)'
     READ(*,*) BR
    END IF
   END IF

   ! TOP_LEFT ELEMENT
   IF (ELEMENT == NUM_ELEMENT_HEAT_LAYER*(NUM_LAYER-1)+1) THEN
    WRITE(*,*)  'Does  TOP_LEFT  ELEMENT  convect  from  two  sides?
(Yes == 1; No == 0)'
    READ(*,*) TWO_SIDE

    IF (TWO_SIDE == 1) THEN
     s = -0.5*BREADTH_HEAT

     kh = 0
     DO i = 1,4
      DO j = 1,4
       DO COUNT_GAUSS = 1, Num_Gauss_Point_TEM
        t = 0.5*(t2-t1)*u(COUNT_GAUSS) + 0.5*(t2+t1)
        CALL SHAPE_FUNCTION_TEM
```

```
       kh1  =  CONVECTION_COEFFI_ELE(ELEMENT)*THICKNESS_TEM*MATMUL
(TRANSPOSE(N),N)
       kh(i,j) = kh(i,j) + 0.5*(t2-t1)*w(COUNT_GAUSS)*kh1(i,j)
      END DO
     END DO
    END DO

    t = 0.5*HEIGHT_HEAT

    DO i = 1,4
     DO j = 1,4
      DO COUNT_GAUSS = 1, Num_Gauss_Point_TEM
       s = 0.5*(s2-s1)*u(COUNT_GAUSS) + 0.5*(s2+s1)
       CALL SHAPE_FUNCTION_TEM
       kh1  =  CONVECTION_COEFFI_ELE(ELEMENT)*THICKNESS_TEM*MATMUL
(TRANSPOSE(N),N)
       kh(i,j) = kh(i,j) + 0.5*(s2-s1)*w(COUNT_GAUSS)*kh1(i,j)
      END DO
     END DO
    END DO
   ELSE
    WRITE(*,*) 'Does the element convect from top side or left
side? (TOP = 1; LEFT = 0)'
    READ(*,*) TL
   END IF
  END IF

  ! TOP_RIGHT ELEMENT
  IF (ELEMENT == NUM_ELEMENT_HEAT) THEN
   WRITE(*,*) 'Does TOP_RIGHT ELEMENT convect from two sides?
(Yes == 1; No == 0)'
   READ(*,*) TWO_SIDE

   IF (TWO_SIDE == 1) THEN
    s = 0.5*BREADTH_HEAT

    kh = 0
    DO i = 1,4
     DO j = 1,4
      DO COUNT_GAUSS = 1, Num_Gauss_Point_TEM
       t = 0.5*(t2-t1)*u(COUNT_GAUSS) + 0.5*(t2+t1)
       CALL SHAPE_FUNCTION_TEM
       kh1  =  CONVECTION_COEFFI_ELE(ELEMENT)*THICKNESS_TEM*MATMUL
(TRANSPOSE(N),N)
       kh(i,j) = kh(i,j) + 0.5*(t2-t1)*w(COUNT_GAUSS)*kh1(i,j)
      END DO
     END DO
    END DO
```

```
      t = 0.5*HEIGHT_HEAT

      DO i = 1,4
       DO j = 1,4
        DO COUNT_GAUSS = 1, Num_Gauss_Point_TEM
         s = 0.5*(s2-s1)*u(COUNT_GAUSS) + 0.5*(s2+s1)
         CALL SHAPE_FUNCTION_TEM
         kh1  =  CONVECTION_COEFFI_ELE(ELEMENT)*THICKNESS_TEM*MATMUL
(TRANSPOSE(N),N)
         kh(i,j) = kh(i,j) + 0.5*(s2-s1)*w(COUNT_GAUSS)*kh1(i,j)
        END DO
       END DO
      END DO
     ELSE
      WRITE(*,*) 'Does the element convect from top side or right
side? (TOP = 1; RIGHT = 0)'
      READ(*,*) TR
     END IF
    END IF

    ! TOP ELEMENTS
    IF(INT((ELEMENT-1)/NUM_ELEMENT_HEAT_LAYER)>=(NUM_LAYER-1))THEN
     IF (ELEMENT == NUM_ELEMENT_HEAT_LAYER*(NUM_LAYER-1)+1) THEN
      IF (TL == 1) THEN
       t = 0.5*HEIGHT_HEAT

       kh = 0
       DO i = 1,4
        DO j = 1,4
         DO COUNT_GAUSS = 1, Num_Gauss_Point_TEM
          s = 0.5*(s2-s1)*u(COUNT_GAUSS) + 0.5*(s2+s1)
          CALL SHAPE_FUNCTION_TEM
          kh1  =  CONVECTION_COEFFI_ELE(ELEMENT)*THICKNESS_TEM*MATMUL
(TRANSPOSE(N),N)
          kh(i,j) = kh(i,j) + 0.5*(s2-s1)*w(COUNT_GAUSS)*kh1(i,j)
         END DO
        END DO
       END DO
      END IF
     END IF

     IF (ELEMENT == NUM_ELEMENT_HEAT) THEN
      IF (TR == 1) THEN
       t = 0.5*HEIGHT_HEAT

       kh = 0
       DO i = 1,4
        DO j = 1,4
         DO COUNT_GAUSS = 1, Num_Gauss_Point_TEM
          s = 0.5*(s2-s1)*u(COUNT_GAUSS) + 0.5*(s2+s1)
          CALL SHAPE_FUNCTION_TEM
```

```
        kh1  =  CONVECTION_COEFFI_ELE(ELEMENT)*THICKNESS_TEM*MATMUL
(TRANSPOSE(N),N)
        kh(i,j) = kh(i,j) + 0.5*(s2-s1)*w(COUNT_GAUSS)*kh1(i,j)
       END DO
      END DO
     END DO
    END IF
   END IF

   IF (ELEMENT /= NUM_ELEMENT_HEAT_LAYER*(NUM_LAYER-1)+1) THEN
    IF (ELEMENT /= NUM_ELEMENT_HEAT) THEN
     t = 0.5*HEIGHT_HEAT

     kh = 0
     DO i = 1,4
      DO j = 1,4
       DO COUNT_GAUSS = 1, Num_Gauss_Point_TEM
        s = 0.5*(s2-s1)*u(COUNT_GAUSS) + 0.5*(s2+s1)
        CALL SHAPE_FUNCTION_TEM
        kh1  =  CONVECTION_COEFFI_ELE(ELEMENT)*THICKNESS_TEM*MATMUL
(TRANSPOSE(N),N)
        kh(i,j) = kh(i,j) + 0.5*(s2-s1)*w(COUNT_GAUSS)*kh1(i,j)
       END DO
      END DO
     END DO
    END IF
   END IF
  END IF

  ! BOTTOM ELEMENTS
  IF (INT((ELEMENT-1)/NUM_ELEMENT_HEAT_LAYER) <= 0) THEN
   IF (ELEMENT == 1) THEN
    IF (BL == 1) THEN
     t = -0.5*HEIGHT_HEAT

     kh = 0
     DO i = 1,4
      DO j = 1,4
       DO COUNT_GAUSS = 1, Num_Gauss_Point_TEM
        s = 0.5*(s2-s1)*u(COUNT_GAUSS) + 0.5*(s2+s1)
        CALL SHAPE_FUNCTION_TEM
        kh1  =  CONVECTION_COEFFI_ELE(ELEMENT)*THICKNESS_TEM*MATMUL
(TRANSPOSE(N),N)
        kh(i,j) = kh(i,j) + 0.5*(s2-s1)*w(COUNT_GAUSS)*kh1(i,j)
       END DO
      END DO
     END DO
    END IF
   END IF
```

```
    IF (ELEMENT == NUM_ELEMENT_HEAT_LAYER) THEN
     IF (BR == 1) THEN
      t = -0.5*HEIGHT_HEAT

      kh = 0
      DO i = 1,4
       DO j = 1,4
        DO COUNT_GAUSS = 1, Num_Gauss_Point_TEM
         s = 0.5*(s2-s1)*u(COUNT_GAUSS) + 0.5*(s2+s1)
         CALL SHAPE_FUNCTION_TEM
         kh1  =  CONVECTION_COEFFI_ELE(ELEMENT)*THICKNESS_TEM*MATMUL
(TRANSPOSE(N),N)
         kh(i,j) = kh(i,j) + 0.5*(s2-s1)*w(COUNT_GAUSS)*kh1(i,j)
        END DO
       END DO
      END DO
     END IF
    END IF

    IF (ELEMENT /= 1) THEN
     IF (ELEMENT /= NUM_ELEMENT_HEAT_LAYER) THEN
      t = -0.5*HEIGHT_HEAT

      kh = 0
      DO i = 1,4
       DO j = 1,4
        DO COUNT_GAUSS = 1, Num_Gauss_Point_TEM
         s = 0.5*(s2-s1)*u(COUNT_GAUSS) + 0.5*(s2+s1)
         CALL SHAPE_FUNCTION_TEM
         kh1  =  CONVECTION_COEFFI_ELE(ELEMENT)*THICKNESS_TEM*MATMUL
(TRANSPOSE(N),N)
         kh(i,j) = kh(i,j) + 0.5*(s2-s1)*w(COUNT_GAUSS)*kh1(i,j)
        END DO
       END DO
      END DO
     END IF
    END IF
   END IF

   ! LEFT ELEMENTS
   LEFT_ELEMENT = 1
   DO WHILE (LEFT_ELEMENT <= NUM_ELEMENT_HEAT)
    IF (ELEMENT == LEFT_ELEMENT) THEN
     IF (ELEMENT == 1) THEN
      IF (BL == 0) THEN
       s = -0.5*BREADTH_HEAT

       kh = 0
       DO i = 1,4
        DO j = 1,4
         DO COUNT_GAUSS = 1, Num_Gauss_Point_TEM
          t = 0.5*(t2-t1)*u(COUNT_GAUSS) + 0.5*(t2+t1)
```

```
          CALL SHAPE_FUNCTION_TEM
          kh1 = CONVECTION_COEFFI_ELE(ELEMENT)*THICKNESS_TEM*MATMUL
(TRANSPOSE(N),N)
          kh(i,j) = kh(i,j) + 0.5*(t2-t1)*w(COUNT_GAUSS)*kh1(i,j)
         END DO
        END DO
       END DO
      END IF
     END IF

     IF (ELEMENT == NUM_ELEMENT_HEAT_LAYER*(NUM_LAYER-1)+1) THEN
      IF (TL == 0) THEN
       s = -0.5*BREADTH_HEAT

       kh = 0
       DO i = 1,4
        DO j = 1,4
         DO COUNT_GAUSS = 1, Num_Gauss_Point_TEM
          t = 0.5*(t2-t1)*u(COUNT_GAUSS) + 0.5*(t2+t1)
          CALL SHAPE_FUNCTION_TEM
          kh1 = CONVECTION_COEFFI_ELE(ELEMENT)*THICKNESS_TEM*MATMUL
(TRANSPOSE(N),N)
          kh(i,j) = kh(i,j) + 0.5*(t2-t1)*w(COUNT_GAUSS)*kh1(i,j)
         END DO
        END DO
       END DO
      END IF
     END IF

     IF (ELEMENT /= 1) THEN
      IF (ELEMENT /= NUM_ELEMENT_HEAT_LAYER*(NUM_LAYER-1)+1) THEN
       s = -0.5*BREADTH_HEAT

       kh = 0
       DO i = 1,4
        DO j = 1,4
         DO COUNT_GAUSS = 1, Num_Gauss_Point_TEM
          t = 0.5*(t2-t1)*u(COUNT_GAUSS) + 0.5*(t2+t1)
          CALL SHAPE_FUNCTION_TEM
          kh1 = CONVECTION_COEFFI_ELE(ELEMENT)*THICKNESS_TEM*MATMUL
(TRANSPOSE(N),N)
          kh(i,j) = kh(i,j) + 0.5*(t2-t1)*w(COUNT_GAUSS)*kh1(i,j)
         END DO
        END DO
       END DO
      END IF
     END IF
    END IF

    LEFT_ELEMENT = LEFT_ELEMENT + NUM_ELEMENT_HEAT_LAYER
   END DO
```

```
 ! RIGHT ELEMENTS
 RIGHT_ELEMENT = NUM_ELEMENT_HEAT_LAYER
 DO WHILE (RIGHT_ELEMENT <= NUM_ELEMENT_HEAT)
  IF (ELEMENT == RIGHT_ELEMENT) THEN
   IF (ELEMENT == NUM_ELEMENT_HEAT_LAYER) THEN
    IF (BR == 0) THEN
     s = 0.5*BREADTH_HEAT

     kh = 0
     DO i = 1,4
      DO j = 1,4
       DO COUNT_GAUSS = 1, Num_Gauss_Point_TEM
        t = 0.5*(t2-t1)*u(COUNT_GAUSS) + 0.5*(t2+t1)
        CALL SHAPE_FUNCTION_TEM
        kh1 = CONVECTION_COEFFI_ELE(ELEMENT)*THICKNESS_TEM*MATMUL
(TRANSPOSE(N),N)
        kh(i,j) = kh(i,j) + 0.5*(t2-t1)*w(COUNT_GAUSS)*kh1(i,j)
       END DO
      END DO
     END DO
    END IF
   END IF

   IF (ELEMENT == NUM_ELEMENT_HEAT) THEN
    IF (TR == 0) THEN
     s = 0.5*BREADTH_HEAT

     kh = 0
     DO i = 1,4
      DO j = 1,4
       DO COUNT_GAUSS = 1, Num_Gauss_Point_TEM
        t = 0.5*(t2-t1)*u(COUNT_GAUSS) + 0.5*(t2+t1)
        CALL SHAPE_FUNCTION_TEM
        kh1 = CONVECTION_COEFFI_ELE(ELEMENT)*THICKNESS_TEM*MATMUL
(TRANSPOSE(N),N)
        kh(i,j) = kh(i,j) + 0.5*(t2-t1)*w(COUNT_GAUSS)*kh1(i,j)
       END DO
      END DO
     END DO
    END IF
   END IF

   IF (ELEMENT /= NUM_ELEMENT_HEAT_LAYER) THEN
    IF (ELEMENT /= NUM_ELEMENT_HEAT) THEN
     s = 0.5*BREADTH_HEAT

     kh = 0
     DO i = 1,4
      DO j = 1,4
       DO COUNT_GAUSS = 1, Num_Gauss_Point_TEM
        t = 0.5*(t2-t1)*u(COUNT_GAUSS) + 0.5*(t2+t1)
```

```
          CALL SHAPE_FUNCTION_TEM
          kh1 = CONVECTION_COEFFI_ELE(ELEMENT)*THICKNESS_TEM*MATMUL
(TRANSPOSE(N),N)
          kh(i,j) = kh(i,j) + 0.5*(t2-t1)*w(COUNT_GAUSS)*kh1(i,j)
         END DO
        END DO
       END DO
      END IF
     END IF
    END IF

    RIGHT_ELEMENT = RIGHT_ELEMENT + NUM_ELEMENT_HEAT_LAYER
   END DO
  END IF
 END DO

END SUBROUTINE ELEMENT_STIFFNESS_HEAT


SUBROUTINE MASS_MATRIX
!-------------------------------------------------------------------------------
! The subroutine MASS_MATRIX generates global
! mass matrix
!-------------------------------------------------------------------------------
 USE MODULE_NAME
 IMPLICIT NONE
 INTEGER :: COUNT_ELEMENT, COUNT_LAYER, COUNT_i, COUNT_j, locate1,
locate2

 locate1 = 0
 locate2 = 0
 MASS_GLOBAL = 0
 DO COUNT_ELEMENT = 1, NUM_ELEMENT_HEAT
  CALL ELEMENT_MASS_MATRIX(COUNT_ELEMENT)

  DO COUNT_LAYER = 1, NUM_LAYER-1
   IF(COUNT_ELEMENT==(COUNT_LAYER*NUM_ELEMENT_HEAT_LAYER+1))THEN
    locate1 = locate1 + 1
    locate2 = locate2 + 1
   END IF
  END DO

  DO COUNT_i = 1,4
   IF (COUNT_i > 2) THEN
    locate1 = locate1 + NUM_ELEMENT_HEAT_LAYER - 1
   END IF

   DO COUNT_j = 1,4
    IF (COUNT_j > 2) THEN
     locate2 = locate2 + NUM_ELEMENT_HEAT_LAYER - 1
    END IF
```

```
    MASS_GLOBAL(COUNT_i+locate1,COUNT_j+locate2)   =   MASS_GLOBAL
(COUNT_i+locate1,COUNT_j+locate2) + mass(COUNT_i,COUNT_j)

    IF (COUNT_j > 2) THEN
     locate2 = locate2 - NUM_ELEMENT_HEAT_LAYER + 1
    END IF
   END DO

   IF (COUNT_i > 2) THEN
    locate1 = locate1 - NUM_ELEMENT_HEAT_LAYER + 1
   END IF
  END DO

  locate1 = locate1 + 1
  locate2 = locate2 + 1

 END DO

END SUBROUTINE MASS_MATRIX


SUBROUTINE ELEMENT_MASS_MATRIX(ELEMENT)
!------------------------------------------------------------------------------------
! The subroutine ELEMENT_MASS_MATRIX
! generates element mass matrix
!------------------------------------------------------------------------------------
 USE MODULE_NAME
 IMPLICIT NONE
 REAL :: s1, s2, t1, t2
 REAL, ALLOCATABLE, DIMENSION(:) :: u, w
 INTEGER :: i, j, COUNT_GAUSS1, COUNT_GAUSS2, ELEMENT

 CALL Gauss_Beam(u,w)

 s1 = -0.5*BREADTH_HEAT
 s2 = 0.5*BREADTH_HEAT
 t1 = -0.5*HEIGHT_HEAT
 t2 = 0.5*HEIGHT_HEAT

 mass = 0
 DO i = 1,4
  DO j = 1,4
   DO COUNT_GAUSS1 = 1, Num_Gauss_Point_TEM
    DO COUNT_GAUSS2 = 1, Num_Gauss_Point_TEM
     s = 0.5*(s2-s1)*u(COUNT_GAUSS2) + 0.5*(s2+s1)
     t = 0.5*(t2-t1)*u(COUNT_GAUSS1) + 0.5*(t2+t1)
     CALL SHAPE_FUNCTION_TEM
     mass1 = SPE_HEAT_ELE(ELEMENT)*MASS_DEN_ELE(ELEMENT)*
THICKNESS_TEM*MATMUL(TRANSPOSE(N),N)
     mass(i,j) = mass(i,j) + 0.5*(t2-t1)*0.5*(s2-s1)*
w(COUNT_GAUSS2)*w(COUNT_GAUSS1)*mass1(i,j)
    END DO
```

```
   END DO
  END DO
 END DO

END SUBROUTINE ELEMENT_MASS_MATRIX


SUBROUTINE TRANSIENT_TEMPERATURE
!------------------------------------------------------------------------------
! The subroutine TRANSIENT_TEMPERATURE
! finds solution for transient heat transfer
!------------------------------------------------------------------------------
 USE MODULE_NAME
 IMPLICIT NONE
 INTEGER :: M, i, j, k, COUNT_SHARE, ii, CONCRETE_ELE
 INTEGER :: COUNT_TEM, COUNT_C
 INTEGER :: BOUNDARY_TIME_MIN
 REAL :: BOUNDARY_TIME, BOUNDARY_TIME_HOUR

 LEFT = (1/TIME_STEP)*MASS_GLOBAL + GAMA*K_TEM_GLOBAL
 RIGHT = (1/TIME_STEP)*MASS_GLOBAL - (1-GAMA)*K_TEM_GLOBAL
 RIGHT_VEC = MATMUL(RIGHT,INITIAL_TEM) + F_TOTAL

 BOUNDARY_TIME_MIN = INT(BOUNDARY_TIME/60)
 BOUNDARY_TIME_HOUR = REAL(BOUNDARY_TIME)/3600

 ! Boundary temperature
 BOUNDARY_NODE_TEM = 0
 DO i = 1, STIFF_DIMENSION
  DO j = 1, NODE_NUM
   IF (i == NODE(j)) THEN

   ! ISO 834
   BOUNDARY_NODE_TEM(i,1)=345*LOG10(8*REAL(BOUNDARY_TIME_MIN)+1)+
AMBIENT_TEM

   ! ASTM E119
   !BOUNDARY_NODE_TEM(i,1)=750*(1-EXP(-3.79553*SQRT
(BOUNDARY_TIME_HOUR)))+170.41*SQRT(BOUNDARY_TIME_HOUR)+
AMBIENT_TEM

   END IF
  END DO
 END DO

 ALLOCATE(RIGHT_ADD(STIFF_DIMENSION,1))

 DO i = 1, STIFF_DIMENSION
  IF (BOUNDARY_NODE_TEM(i,1) /= 0) THEN
   RIGHT_VEC(i,1) = BOUNDARY_NODE_TEM(i,1)
   RIGHT_ADD = 0
   DO j = 1, STIFF_DIMENSION
```

```
    IF (j == i) THEN
     DO k = 1, STIFF_DIMENSION
      DO COUNT_SHARE = 1, 12
       IF (k == NODE_SHARE(j,COUNT_SHARE)) THEN
        RIGHT_ADD(k,1) = RIGHT_ADD(k,1) - LEFT(j,k)*RIGHT_VEC(j,1)
       END IF
      END DO
     END DO
    END IF
   END DO

   DO j = 1, STIFF_DIMENSION
    IF (j == i) THEN
     DO k = 1, STIFF_DIMENSION
      LEFT(j,k) = 0
     END DO
    END IF
   END DO

   DO k = 1, STIFF_DIMENSION
    IF (k == i) THEN
     DO j = 1, STIFF_DIMENSION
      LEFT(j,k) = 0
     END DO
    END IF
   END DO

   LEFT(i,i) = 1
   RIGHT_VEC = RIGHT_VEC + RIGHT_ADD
  END IF
 END DO

 M = STIFF_DIMENSION + 1
 Num_Unknown_TEM = STIFF_DIMENSION

 ALLOCATE(AUG(STIFF_DIMENSION,M))

 AUG = 0
 DO i = 1, STIFF_DIMENSION
  DO j = 1, STIFF_DIMENSION
   AUG(i,j) = LEFT(i,j)
  END DO
 END DO

 DO i = 1, STIFF_DIMENSION
  AUG(i,M) = RIGHT_VEC(i,1)
 END DO

 CALL GAUSS_ELIMINATION_TEM
 IF (SOLVBL) THEN
  CALL BACK_TEM(NODAL_TEM)
 ELSE
```

```
 WRITE(*,*) 'NO SOLUTION EXISTS'
END IF

! FOR CONCRETE
COUNT_TEM = 0
TEM = 0
DO i = 1, Num_Layer_Material(1)
 DO j = 1, NUM_TEM_LAYER(1)
  COUNT_TEM = COUNT_TEM + 1
  ii = COUNT_TEM + i - 1
  DO k = 1,4
   IF (k > 2) THEN
    ii = ii + NUM_ELEMENT_HEAT_LAYER - 1
   END IF

   TEM(1,COUNT_TEM) = TEM(1,COUNT_TEM) + 0.25*NODAL_TEM(ii,1)

   IF (k > 2) THEN
    ii = ii - NUM_ELEMENT_HEAT_LAYER + 1
   END IF
   ii = ii + 1
  END DO
 END DO
END DO

! FOR REINFORCING BARS
IF (Num_Material_Ele > 1) THEN
 DO i = 2, Num_Material_Ele
  COUNT_TEM = 0
  DO j = 1, Num_Layer_Material(i)
   DO k = 1, NUM_TEM_LAYER(i)
    COUNT_TEM = COUNT_TEM + 1
    DO COUNT_C = 1, Num_Layer_Material(1)
     IF ((Material_Top(i,j) + Material_Bot(i,j))/2 >= Material_Bot
(1,COUNT_C)) THEN
      IF ((Material_Top(i,j) + Material_Bot(i,j))/2 <= Material_Top
(1,COUNT_C)) THEN
       CONCRETE_ELE = (COUNT_C-1)*NUM_ELEMENT_HEAT_LAYER + INT
(DIS_2(i,COUNT_TEM)/BREADTH_HEAT)+1
       TEM(i,COUNT_TEM) = TEM(1,CONCRETE_ELE)
      END IF
     END IF
    END DO
   END DO
  END DO
 END DO
ENDIF

DEALLOCATE(RIGHT_ADD)
DEALLOCATE(AUG)

END SUBROUTINE TRANSIENT_TEMPERATURE
```

```
SUBROUTINE TEMPERATURE_MATERIAL
!-------------------------------------------------------------------------------
! The subroutine TEMPERATURE_MATERIAL
! finds temperature dependent coefficients of
! material properties
!-------------------------------------------------------------------------------
 USE MODULE_NAME
 IMPLICIT NONE
 REAL :: TOT, TOT_STRAIN, TOT_STRAIN_ce, TOT_E, TOT_ap, TOT_au
 INTEGER :: i, j, k, m, COUNT_TEM, MAX

 E_REDU = 0

 MAX = Num_Layer_Material(1)
 IF (Num_Material_Ele > 1) THEN
  DO i = 2, Num_Material_Ele
   IF (MAX < Num_Layer_Material(i)) THEN
    MAX = Num_Layer_Material(i)
   END IF
  END DO
 END IF

 ALLOCATE(k_AVERAGE(Num_Material_Ele,MAX),E_AVERAGE
(Num_Material_Ele,MAX))
 ALLOCATE(k_ap_AVERAGE(Num_Material_Ele,MAX),k_au_AVERAGE
(Num_Material_Ele,MAX))
 ALLOCATE(STRAIN_cu_AVERAGE(Num_Material_Ele,MAX))
 ALLOCATE (STRAIN_ce_AVERAGE(Num_Material_Ele,MAX))
 ALLOCATE(Db_TEM(Num_Material_Ele,MAX),Ds_TEM(Num_Material_Ele,MAX))
 ALLOCATE(Crack_Stress_TEM(Num_Material_Ele,MAX),Ulti_Stress_TEM
(Num_Material_Ele,MAX))
 ALLOCATE(STRESS_ap(Num_Material_Ele,MAX),STRESS_au
(Num_Material_Ele,MAX))
 ALLOCATE(Crack_Strain_TEM(Num_Material_Ele,MAX),
Tension_Strain_TEM(Num_Material_Ele,MAX))

 ! CONCRETE (IN COMPRESSION)
 IF (CONCRETE_TYPE == 1) THEN ! (1-siliceous; 2-calcareous)
 COUNT_TEM = 0
  DO i = 1, Num_Layer_Material(1)
   DO j = 1, NUM_TEM_LAYER(1)
    COUNT_TEM = COUNT_TEM + 1
    IF (TEM(1,COUNT_TEM) <= 100) THEN
     k_REDU(1,COUNT_TEM) = 1.0
    ELSE IF (TEM(1,COUNT_TEM) <= 400) THEN
     k_REDU(1,COUNT_TEM) = 1.067-0.00067*TEM(1,COUNT_TEM)
    ELSE IF (TEM(1,COUNT_TEM) <= 900) THEN
     k_REDU(1,COUNT_TEM) = 1.44-0.0016*TEM(1,COUNT_TEM)
    ELSE
     k_REDU(1,COUNT_TEM) = 0
```

```
   END IF
  END DO
 END DO
END IF

IF (CONCRETE_TYPE == 2) THEN
COUNT_TEM = 0
 DO i = 1, Num_Layer_Material(1)
  DO j = 1, NUM_TEM_LAYER(1)
   COUNT_TEM = COUNT_TEM + 1
   IF (TEM(1,COUNT_TEM) <= 20) THEN
    k_REDU(1,COUNT_TEM) = 1.0
   ELSE IF (TEM(1,COUNT_TEM) <= 100) THEN
    k_REDU(1,COUNT_TEM) = 1.0075-0.000375*TEM(1,COUNT_TEM)
   ELSE IF (TEM(1,COUNT_TEM) <= 300) THEN
    k_REDU(1,COUNT_TEM) = 1.0-0.0003*TEM(1,COUNT_TEM)
   ELSE IF (TEM(1,COUNT_TEM) <= 400) THEN
    k_REDU(1,COUNT_TEM) = 1.09-0.0006*TEM(1,COUNT_TEM)
   ELSE IF (TEM(1,COUNT_TEM) <= 500) THEN
    k_REDU(1,COUNT_TEM) = 1.29-0.0011*TEM(1,COUNT_TEM)
   ELSE IF (TEM(1,COUNT_TEM) <= 600) THEN
    k_REDU(1,COUNT_TEM) = 1.44-0.0014*TEM(1,COUNT_TEM)
   ELSE IF (TEM(1,COUNT_TEM) <= 700) THEN
    k_REDU(1,COUNT_TEM) = 1.62-0.0017*TEM(1,COUNT_TEM)
   ELSE IF (TEM(1,COUNT_TEM) <= 800) THEN
    k_REDU(1,COUNT_TEM) = 1.55-0.0016*TEM(1,COUNT_TEM)
   ELSE IF (TEM(1,COUNT_TEM) <= 900) THEN
    k_REDU(1,COUNT_TEM) = 1.23-0.0012*TEM(1,COUNT_TEM)
   ELSE IF (TEM(1,COUNT_TEM) <= 1000) THEN
    k_REDU(1,COUNT_TEM) = 0.96-0.0009*TEM(1,COUNT_TEM)
   ELSE IF (TEM(1,COUNT_TEM) <= 1100) THEN
    k_REDU(1,COUNT_TEM) = 0.46-0.0004*TEM(1,COUNT_TEM)
   ELSE
    k_REDU(1,COUNT_TEM) = 0
   END IF
  END DO
 END DO
END IF

COUNT_TEM = 0
DO i = 1, Num_Layer_Material(1)
 DO j = 1, NUM_TEM_LAYER(1)
  COUNT_TEM = COUNT_TEM + 1
  IF (TEM(1,COUNT_TEM) <= 20) THEN
   STRAIN_cu(1,COUNT_TEM) = 0.0025
  ELSE IF (TEM(1,COUNT_TEM) <= 100) THEN
   STRAIN_cu(1,COUNT_TEM)=0.0025+0.00001875*(TEM(1,COUNT_TEM)-20)
  ELSE IF (TEM(1,COUNT_TEM) <= 300) THEN
   STRAIN_cu(1,COUNT_TEM) = 0.004+0.000015*(TEM(1,COUNT_TEM)-100)
```

```
    ELSE IF (TEM(1,COUNT_TEM) <= 400) THEN
     STRAIN_cu(1,COUNT_TEM) = 0.007+0.00003*(TEM(1,COUNT_TEM)-300)
    ELSE IF (TEM(1,COUNT_TEM) <= 500) THEN
     STRAIN_cu(1,COUNT_TEM) = 0.01+0.00005*(TEM(1,COUNT_TEM)-400)
    ELSE IF (TEM(1,COUNT_TEM) <= 600) THEN
     STRAIN_cu(1,COUNT_TEM) = 0.015+0.0001*(TEM(1,COUNT_TEM)-500)
    ELSE
     STRAIN_cu(1,COUNT_TEM) = 0.025
    END IF

    IF (TEM(1,COUNT_TEM) <= 20) THEN
     STRAIN_ce(1,COUNT_TEM) = 0.02
    ELSE IF (TEM(1,COUNT_TEM) <= 100) THEN
     STRAIN_ce(1,COUNT_TEM) = 0.02 + 0.00003125*(TEM(1,COUNT_TEM)-20)
    ELSE IF (TEM(1,COUNT_TEM) <= 1100) THEN
     STRAIN_ce(1,COUNT_TEM)=0.0225+0.000025*(TEM(1,COUNT_TEM)-100)
    ELSE
     STRAIN_ce(1,COUNT_TEM) = 0.0475
    END IF

    IF (TEM(1,COUNT_TEM) <= 100) THEN
     SPE_HEAT_ELE(COUNT_TEM) = 900
    ELSE IF (TEM(1,COUNT_TEM) <= 200) THEN
     SPE_HEAT_ELE(COUNT_TEM) = 900+(TEM(1,COUNT_TEM)-100)
    ELSE IF (TEM(1,COUNT_TEM) <= 400) THEN
     SPE_HEAT_ELE(COUNT_TEM) = 1000+(TEM(1,COUNT_TEM)-200)/2
    ELSE
     SPE_HEAT_ELE(COUNT_TEM) = 1100
    END IF

    THERMAL_CONDUCT_ELE_x(COUNT_TEM)  =  2-0.2451*(TEM(1,COUNT_TEM)/
 100)+0.0107*(TEM(1,COUNT_TEM)/100)**2
    THERMAL_CONDUCT_ELE_y(COUNT_TEM)  =  2-0.2451*(TEM(1,COUNT_TEM)/
 100)+0.0107*(TEM(1,COUNT_TEM)/100)**2
    MASS_DEN_ELE(COUNT_TEM) = 2354-23.47*(TEM(1,COUNT_TEM)/100)
   END DO
  END DO

  COUNT_TEM = 0
  DO i = 1, Num_Layer_Material(1)
   TOT = 0
   TOT_STRAIN = 0
   TOT_STRAIN_ce = 0
   DO j = 1, NUM_TEM_LAYER(1)
    COUNT_TEM = COUNT_TEM + 1
    TOT = TOT + k_REDU(1,COUNT_TEM)
    TOT_STRAIN = TOT_STRAIN + STRAIN_cu(1,COUNT_TEM)
    TOT_STRAIN_ce = TOT_STRAIN_ce + STRAIN_ce(1,COUNT_TEM)
   END DO
```

```
 k_AVERAGE(1,i) = TOT/REAL(NUM_TEM_LAYER(1))
 STRAIN_cu_AVERAGE(1,i) = TOT_STRAIN/REAL(NUM_TEM_LAYER(1))
 STRAIN_ce_AVERAGE(1,i) = TOT_STRAIN_ce/REAL(NUM_TEM_LAYER(1))
 Ulti_Stress_TEM(1,i) = k_AVERAGE(1,i)*Ulti_Stress(1)
 Crack_Stress_TEM(1,i) = 0.33*SQRT(Ulti_Stress_TEM(1,i))
 Db_TEM(1,i) = 4700*SQRT(Ulti_Stress_TEM(1,i))
 Ds_TEM(1,i) = Db_TEM(1,i)/(2*(1+POISSON))

 Crack_Strain_TEM(1,i) = Crack_Stress_TEM(1,i)/Db_TEM(1,i)
END DO

! REINFORCING BARS (TENSILE STRENGTH & YOUNG'S MODULUS)
AREA_TOTAL_TEM = 0
EA_TOTAL_TEM = 0
STRAIN_A_TOTAL_TEM = 0
IF (Num_Material_Ele > 1) THEN
 DO k = 2, Num_Material_Ele
  IF (REO_TYPE(k) == 1) THEN
   ! STEEL
   COUNT_TEM = 0
   DO i = 1, Num_Layer_Material(k)
    DO j = 1, NUM_TEM_LAYER(k)
     COUNT_TEM = COUNT_TEM + 1

     ! reduction factor for yield strength f_ay (EUROCODE 4)
     IF (TEM(k,COUNT_TEM) <= 400) THEN
      k_REDU(k,COUNT_TEM) = 1.0
     ELSE IF (TEM(k,COUNT_TEM) <= 500) THEN
      k_REDU(k,COUNT_TEM) = 1.88-0.0022*TEM(k,COUNT_TEM)
     ELSE IF (TEM(k,COUNT_TEM) <= 600) THEN
      k_REDU(k,COUNT_TEM) = 2.33-0.0031*TEM(k,COUNT_TEM)
     ELSE IF (TEM(k,COUNT_TEM) <= 700) THEN
      k_REDU(k,COUNT_TEM) = 1.91-0.0024*TEM(k,COUNT_TEM)
     ELSE IF (TEM(k,COUNT_TEM) <= 800) THEN
      k_REDU(k,COUNT_TEM) = 1.07-0.0012*TEM(k,COUNT_TEM)
     ELSE IF (TEM(k,COUNT_TEM) <= 900) THEN
      k_REDU(k,COUNT_TEM) = 0.51-0.0005*TEM(k,COUNT_TEM)
     ELSE IF (TEM(k,COUNT_TEM) <= 1100) THEN
      k_REDU(k,COUNT_TEM) = 0.24-0.0002*TEM(k,COUNT_TEM)
     ELSE
      k_REDU(k,COUNT_TEM) = 0
     END IF

     ! reduction factor for proportional limit f_ap (EUROCODE 4)
     IF (TEM(k,COUNT_TEM) <= 100) THEN
      k_ap(k,COUNT_TEM) = 1.0
     ELSE IF (TEM(k,COUNT_TEM) <= 200) THEN
      k_ap(k,COUNT_TEM) = 1.193-0.00193*TEM(k,COUNT_TEM)
     ELSE IF (TEM(k,COUNT_TEM) <= 300) THEN
      k_ap(k,COUNT_TEM) = 1.195-0.00194*TEM(k,COUNT_TEM)
     ELSE IF (TEM(k,COUNT_TEM) <= 400) THEN
```

```
 k_ap(k,COUNT_TEM) = 1.192-0.00193*TEM(k,COUNT_TEM)
ELSE IF (TEM(k,COUNT_TEM) <= 500) THEN
 k_ap(k,COUNT_TEM) = 0.66-0.0006*TEM(k,COUNT_TEM)
ELSE IF (TEM(k,COUNT_TEM) <= 600) THEN
 k_ap(k,COUNT_TEM) = 1.26-0.0018*TEM(k,COUNT_TEM)
ELSE IF (TEM(k,COUNT_TEM) <= 700) THEN
 k_ap(k,COUNT_TEM) = 0.81-0.00105*TEM(k,COUNT_TEM)
ELSE IF (TEM(k,COUNT_TEM) <= 800) THEN
 k_ap(k,COUNT_TEM) = 0.25-0.00025*TEM(k,COUNT_TEM)
ELSE IF (TEM(k,COUNT_TEM) <= 1100) THEN
 k_ap(k,COUNT_TEM) = 0.15-0.000125*TEM(k,COUNT_TEM)
ELSE
 k_ap(k,COUNT_TEM) = 0
END IF

! reduction factor for f_au (EUROCODE 4) (USED FOR STRAIN HARDENING)
IF (TEM(k,COUNT_TEM) <= 300) THEN
 k_au(k,COUNT_TEM) = 1.25
ELSE IF (TEM(k,COUNT_TEM) <= 400) THEN
 k_au(k,COUNT_TEM) = 2-0.0025*TEM(k,COUNT_TEM)
ELSE IF (TEM(k,COUNT_TEM) <= 500) THEN
 k_au(k,COUNT_TEM) = 1.88-0.0022*TEM(k,COUNT_TEM)
ELSE IF (TEM(k,COUNT_TEM) <= 600) THEN
 k_au(k,COUNT_TEM) = 2.33-0.0031*TEM(k,COUNT_TEM)
ELSE IF (TEM(k,COUNT_TEM) <= 700) THEN
 k_au(k,COUNT_TEM) = 1.91-0.0024*TEM(k,COUNT_TEM)
ELSE IF (TEM(k,COUNT_TEM) <= 800) THEN
 k_au(k,COUNT_TEM) = 1.07-0.0012*TEM(k,COUNT_TEM)
ELSE IF (TEM(k,COUNT_TEM) <= 900) THEN
 k_au(k,COUNT_TEM) = 0.51-0.0005*TEM(k,COUNT_TEM)
ELSE IF (TEM(k,COUNT_TEM) <= 1100) THEN
 k_au(k,COUNT_TEM) = 0.24-0.0002*TEM(k,COUNT_TEM)
ELSE
 k_au(k,COUNT_TEM) = 0
END IF

IF (TEM(k,COUNT_TEM) <= 100) THEN
 E_REDU(k,COUNT_TEM) = 1.0
ELSE IF (TEM(k,COUNT_TEM) <= 500) THEN
 E_REDU(k,COUNT_TEM) = 1.10-0.001*TEM(k,COUNT_TEM)
ELSE IF (TEM(k,COUNT_TEM) <= 600) THEN
 E_REDU(k,COUNT_TEM) = 2.05-0.0029*TEM(k,COUNT_TEM)
ELSE IF (TEM(k,COUNT_TEM) <= 700) THEN
 E_REDU(k,COUNT_TEM) = 1.39-0.0018*TEM(k,COUNT_TEM)
ELSE IF (TEM(k,COUNT_TEM) <= 800) THEN
 E_REDU(k,COUNT_TEM) = 0.41-0.0004*TEM(k,COUNT_TEM)
ELSE IF (TEM(k,COUNT_TEM) <= 1200) THEN
 E_REDU(k,COUNT_TEM) = 0.27-0.000225*TEM(k,COUNT_TEM)
ELSE
```

```
      E_REDU(k,COUNT_TEM) = 0
     END IF
    END DO
   END DO

  ELSE IF (REO_TYPE(k) == 2) THEN
   ! GFRP
   COUNT_TEM = 0
   DO i = 1, Num_Layer_Material(k)
    DO j = 1, NUM_TEM_LAYER(k)
     COUNT_TEM = COUNT_TEM + 1

     IF (TEM(k,COUNT_TEM) <= 400) THEN
      k_REDU(k,COUNT_TEM) = 1-0.0025*TEM(k,COUNT_TEM)
     ELSE
      k_REDU(k,COUNT_TEM) = 0
     END IF

     IF (TEM(k,COUNT_TEM) <= 100) THEN
      E_REDU(k,COUNT_TEM) = 1.0
     ELSE IF (TEM(k,COUNT_TEM) <= 300) THEN
      E_REDU(k,COUNT_TEM) = 1.25-0.0025*TEM(k,COUNT_TEM)
     ELSE IF (TEM(k,COUNT_TEM) <= 400) THEN
      E_REDU(k,COUNT_TEM) = 2.0-0.005*TEM(k,COUNT_TEM)
     ELSE
      E_REDU(k,COUNT_TEM) = 0
     END IF
    END DO
   END DO

  ELSE IF (REO_TYPE(k) == 3) THEN
   ! AFRP
   COUNT_TEM = 0
   DO i = 1, Num_Layer_Material(k)
    DO j = 1, NUM_TEM_LAYER(k)
     COUNT_TEM = COUNT_TEM + 1

     IF (TEM(k,COUNT_TEM) <= 100) THEN
      k_REDU(k,COUNT_TEM) = 1
     ELSE IF (TEM(k,COUNT_TEM) <= 400) THEN
      k_REDU(k,COUNT_TEM) = 1.333-0.00333*TEM(k,COUNT_TEM)
     ELSE
      k_REDU(k,COUNT_TEM) = 0
     END IF

     IF (TEM(k,COUNT_TEM) <= 100) THEN
      E_REDU(k,COUNT_TEM) = 1.0
     ELSE IF (TEM(k,COUNT_TEM) <= 300) THEN
      E_REDU(k,COUNT_TEM) = 1.25-0.0025*TEM(k,COUNT_TEM)
     ELSE IF (TEM(k,COUNT_TEM) <= 400) THEN
      E_REDU(k,COUNT_TEM) = 2.0-0.005*TEM(k,COUNT_TEM)
```

```
      ELSE
       E_REDU(k,COUNT_TEM) = 0
      END IF
     END DO
    END DO

   ELSE IF (REO_TYPE(k) == 4) THEN
    ! CFRP
    COUNT_TEM = 0
    DO i = 1, Num_Layer_Material(k)
     DO j = 1, NUM_TEM_LAYER(k)
      COUNT_TEM = COUNT_TEM + 1

      IF (TEM(k,COUNT_TEM) <= 200) THEN
       k_REDU(k,COUNT_TEM) = 1
      ELSE IF (TEM(k,COUNT_TEM) < 725) THEN
       k_REDU(k,COUNT_TEM) = 1.380952381-0.001904761905*TEM
(k,COUNT_TEM)
      ELSE
       k_REDU(k,COUNT_TEM) = 0
      END IF

      IF (TEM(k,COUNT_TEM) <= 200) THEN
       E_REDU(k,COUNT_TEM) = 1.0
      ELSE IF (TEM(k,COUNT_TEM) <= 450) THEN
       E_REDU(k,COUNT_TEM) = 1.28-0.0014*TEM(k,COUNT_TEM)
      ELSE IF (TEM(k,COUNT_TEM) < 750) THEN
       E_REDU(k,COUNT_TEM) = 1.625-0.00216667*TEM(k,COUNT_TEM)
      ELSE
       E_REDU(k,COUNT_TEM) = 0
      END IF
     END DO
    END DO
   END IF

   COUNT_TEM = 0
   DO i = 1, Num_Layer_Material(k)
    TOT = 0
    TOT_E = 0
    TOT_ap = 0
    TOT_au = 0
    DO j = 1, NUM_TEM_LAYER(k)
     COUNT_TEM = COUNT_TEM + 1
     TOT = TOT + k_REDU(k,COUNT_TEM)
     TOT_E = TOT_E + E_REDU(k,COUNT_TEM)
     TOT_ap = TOT_ap + k_ap(k,COUNT_TEM)
     TOT_au = TOT_au + k_au(k,COUNT_TEM)
    END DO
    k_AVERAGE(k,i) = TOT/REAL(NUM_TEM_LAYER(k))
    E_AVERAGE(k,i) = TOT_E/REAL(NUM_TEM_LAYER(k))
```

```
    k_ap_AVERAGE(k,i) = TOT_ap/REAL(NUM_TEM_LAYER(k))
    k_au_AVERAGE(k,i) = TOT_au/REAL(NUM_TEM_LAYER(k))

    Ulti_Stress_TEM(k,i) = k_AVERAGE(k,i)*Ulti_Stress(k)

    IF (REO_TYPE(k) == 1) THEN
     Crack_Stress_TEM(k,i) = 0
     STRESS_ap(k,i) = k_ap_AVERAGE(k,i)*Ulti_Stress(k)
     STRESS_au(k,i) = k_au_AVERAGE(k,i)*Ulti_Stress(k)
    ELSE
     Crack_Stress_TEM(k,i) = Ulti_Stress_TEM(k,i)
    END IF

    Db_TEM(k,i) = E_AVERAGE(k,i)*Db(k)
    Ds_TEM(k,i) = 0

    Tension_Strain_TEM(k,i) = Ulti_Stress_TEM(k,i)/Db_TEM(k,i)
   END DO

   IF (AREA(k) > 0) THEN
    DO i = 1, Num_Layer_Material(k)
     AREA_TOTAL_TEM = AREA_TOTAL_TEM + breadth*(Material_Top(k,i)-
Material_Bot(k,i))
     EA_TOTAL_TEM = EA_TOTAL_TEM + Db_TEM(k,i)*breadth*
(Material_Top(k,i)-Material_Bot(k,i))
     STRAIN_A_TOTAL_TEM = STRAIN_A_TOTAL_TEM + Tension_Strain_TEM
(k,i)*breadth*(Material_Top(k,i)-Material_Bot(k,i))
    END DO
   END IF
  END DO

  STRAIN_BAR_TEM = STRAIN_A_TOTAL_TEM/AREA_TOTAL_TEM

 END IF

END SUBROUTINE TEMPERATURE_MATERIAL


SUBROUTINE THERMAL_STRAIN
!-----------------------------------------------------------------------------
! The subroutine THERMAL_STRAIN calculates
! thermal strains based on temperature variation
!-----------------------------------------------------------------------------
 USE MODULE_NAME
 IMPLICIT NONE
 REAL :: TOT_STRAIN_TH
 INTEGER :: MAX, i, j, k, COUNT_TEM

 ! CONCRETE
 IF (CONCRETE_TYPE == 1) THEN
  COUNT_TEM = 0
  DO i = 1, Num_Layer_Material(1)
```

```
    DO j = 1, NUM_TEM_LAYER(1)
     COUNT_TEM = COUNT_TEM + 1
     IF (TEM(1,COUNT_TEM) <= 700) THEN
      STRAIN_TH(1,COUNT_TEM)=-0.00018+0.000009*TEM(1,COUNT_TEM)+
0.000000000023*TEM(1,COUNT_TEM)**3
     ELSE IF (TEM(1,COUNT_TEM) <= 1200) THEN
      STRAIN_TH(1,COUNT_TEM) = 0.014
     END IF
    END DO
   END DO
  ELSE IF (CONCRETE_TYPE == 2) THEN
   COUNT_TEM = 0
   DO i = 1, Num_Layer_Material(1)
    DO j = 1, NUM_TEM_LAYER(1)
     COUNT_TEM = COUNT_TEM + 1
     IF (TEM(1,COUNT_TEM) <= 805) THEN
      STRAIN_TH(1,COUNT_TEM)=-0.00012+0.000006*TEM(1,COUNT_TEM)+
0.000000000014*TEM(1,COUNT_TEM)**3
     ELSE IF (TEM(1,COUNT_TEM) <= 1200) THEN
      STRAIN_TH(1,COUNT_TEM) = 0.012
     END IF
    END DO
   END DO
  END IF

  COUNT_TEM = 0
  DO i = 1, Num_Layer_Material(1)
   TOT_STRAIN_TH = 0
   DO j = 1, NUM_TEM_LAYER(1)
    COUNT_TEM = COUNT_TEM + 1
    TOT_STRAIN_TH = TOT_STRAIN_TH + STRAIN_TH(1,COUNT_TEM)
   END DO
   STRAIN_TH_AVERAGE(1,i) = TOT_STRAIN_TH/REAL(NUM_TEM_LAYER(1))
  END DO

  ! REINFORCING BARS
  IF (Num_Material_Ele > 1) THEN
   DO k = 2, Num_Material_Ele
    COUNT_TEM = 0
    DO i = 1, Num_Layer_Material(k)
     DO j = 1, NUM_TEM_LAYER(k)
      COUNT_TEM = COUNT_TEM + 1
      IF (REO_TYPE(k) == 1) THEN
       IF (TEM(k,COUNT_TEM) <= 750) THEN
        STRAIN_TH(k,COUNT_TEM) = -0.0002416 + 0.000012*TEM
(k,COUNT_TEM) + 0.000000004*TEM(k,COUNT_TEM)**2
       ELSE IF (TEM(k,COUNT_TEM) <= 860) THEN
        STRAIN_TH(k,COUNT_TEM) = 0.011
       ELSE
```

```
      STRAIN_TH(k,COUNT_TEM) = -0.0062 + 0.00002*TEM(k,COUNT_TEM)
     END IF
    ELSE IF (REO_TYPE(k) == 2) THEN
     STRAIN_TH(k,COUNT_TEM) = (TEM(k,COUNT_TEM)-AMBIENT_TEM)*
8/1000000
    ELSE IF (REO_TYPE(k) == 3) THEN
     STRAIN_TH(k,COUNT_TEM) = (TEM(k,COUNT_TEM)-AMBIENT_TEM)*(-4)/
1000000
    ELSE IF (REO_TYPE(k) == 4) THEN
     STRAIN_TH(k,COUNT_TEM) = 0
    END IF
   END DO
  END DO

  COUNT_TEM = 0
  DO i = 1, Num_Layer_Material(k)
   TOT_STRAIN_TH = 0
   DO j = 1, NUM_TEM_LAYER(k)
    COUNT_TEM = COUNT_TEM + 1
    TOT_STRAIN_TH = TOT_STRAIN_TH + STRAIN_TH(k,COUNT_TEM)
   END DO
   STRAIN_TH_AVERAGE(k,i) = TOT_STRAIN_TH/REAL(NUM_TEM_LAYER(k))
  END DO
 END DO
END IF

END SUBROUTINE THERMAL_STRAIN
```

Appendix J: User subroutine for concrete under cyclic load

This appendix provides the finite element code for the nonlinear analysis of concrete under cyclic load, which is employed as a user-defined subroutine in ANSYS.

```
*deck,usercr USERDISTRIB parallel
subroutine usercr (elem,intpt,mat,ncomp,kfirst,kfsteq,e,posn,d,
x proptb,timval,timinc,tem,dtem,toffst,fluen,dfluen,epel,epcrp,
x statev,usvr,delcr)
c written by prabin August 2015
c numerical analysis of FRP strengthened RC beams under cyclic load
c *** primary function: allow users to write their own creep laws.
c this logic is accessed with c6 = 100
c *** secondary function: demonstrate the use of user-written creep laws
c *** Notice - This file contains ANSYS Confidential information ***

c *** Copyright ANSYS. All Rights Reserved.
c *** ansys, inc.

c input arguments:
c variable (type,sze,intent) description

c elem (int,sc,in) - element number (label)
c intpt (int,sc,in) - element integration point number
c mat (int,sc,in) - material reference number
c ncomp (int,sc,in) - no. of stress/strain components (1,4 or 6)
c 1 - x
c 4 - x,y,z,xy
c 6 - x,y,z,xy,yz,xz
c kfirst (int,sc,in) - 1 if first time through, 0
otherwise
c (useful for initializing state variables
c to a non-zero value)
c kfsteq (int,sc,in) - 1 if first equilibrium iteration of a
c substep, 0 otherwise

c e (dp,sc,in) - elastic young'S MODULUS
c posn (dp,sc,in) - poisson'S RATIO
c d (dp,ar(ncomp,ncomp),in)- elastic stress-strain matrix
c proptb (dp,ar(72),in) - material properties input on tb commands
c (do not use proptb(13), as it is used elsewhere)
c timval (dp,sc,in) - current time value
```

```
c timinc (dp,sc,in) - time increment over this substep
c tem (dp,sc,in) - temperature at the end of this substep
c dtem (dp,sc,in) - temperature increment over this substep
c toffst (dp,sc,in) - temperature offset from absolute zero
c fluen (dp,sc,in) - fluence at the end of this substep
c dfluen (dp,sc,in) - fluence increment over this substep

c epel (dp,ar(ncomp),inout)- elastic strain
c epcrp (dp,ar(ncomp),inout)- creep strain from previous substep
c statev (dp,ar(ncomp*5+2),inout)- state variables from previous
c (converged) substep. This variable is for
c explicit creep only and refers to a different internal variable
than that
c defined by TB,stat which is used by
c implicit creep (usercreep) and usermat.
c usvr (dp,ar(nuval,nintp),inout)- additional state variables from
c previous equilibrium iteration
(saved if the nsvr command is used)
c
c
c output arguments:
c variable (type,sze,intent) description
c
c epel (dp,ar(ncomp),inout)- elastic strain adjusted for creep
increment
c epcrp (dp,ar(ncomp),inout)- updated creep strain

c statev (dp,ar(ncomp*5+2),inout)- updated state variables
c usvr (dp,ar(nuval,nintp),inout)- updated additional state variables
c delcr (dp,sc,out) - equivalent creep strain increment
(used for creep ratio calculation)

c fortran parameters (to be defined by the user):
c variable (type) description
c nuval (int) - number of additional state variables per
c integration point
c nintp (int) - maximum number of integration points of
c an element to be used with this routine
c (14 is the maximum)
c note: nuval x nintp = nstv(on nsvr command); cannot exceed 840!
c
c internal variables:
c variable (type,sze) description
c con (dp,sc) - temporary variable
c del (dp,ar(6)) - creep strain increments
c epet (dp,sc) - equivalent elastic strain (before creep)
c ept (dp,ar(6)) - total strain
c eptot (dp,sc) - equivalent total strain, elastic + creep
c sigen (dp,sc) - equivalent stress (before creep)
c temabs (dp,sc) - temperature on the absolute scale
```

```
c
#include "impcom.inc"
external erhandler
c
c user-defined fortran parameters
c - size of usvr data
integer nuval,nintp
parameter (nuval=1,nintp=1)
c
c external subroutines and functions
external egen
double precision egen
external vapb,vapb1,vamb1,vmult
c
c integer variables
integer elem,intpt,mat,ncomp,kfirst,kfsteq
c
c double precision variables ——added by prabin
double precision
x e,posn,d(ncomp,ncomp),proptb(72),timval,timinc,tem,dtem,toffst,
x fluen,dfluen,epel(ncomp),epcrp(ncomp),statev(ncomp*5+2),
x usvr(nuval,nintp),delcr,
x temabs,con,del(6),epet,sigen,eptot,ept(6),
x en,para1,para2,smax,nfinal,fre
c
c **** the current version of the ansys program does not accumulate
c the effects of creep work. if this is desired, it is recommened to:
c 1. compute the desired creep work in this routine.
c this can be accumulated over the element. this can
c be done by: if (intpt .eq. 1) usvr(1,1) = 0.0 at
c the first of this routine. then, later, accumulate
c over all integration points as:
c usvr(1,1) = usvr(1,1) + xxxxx
c note that this is really only a creep work density,
c and, if the total creep work is desired, will need
c to be multiplied by the volume as indicated in the
c next steps.
c actually, it is a bit more complex than this. We want
c this accumulation to be done only for converged
c equilibrium iterations. the concept that may be used is:
c if (kfstps .eq. 1) work = 0.0
c if (kfsteq .eq. 1) workold = work
c work = workold + delta(work)
c thus, at the converged equilibrium iteration,
c work will be correct.
c 2. in userou, move the data from usvr(1,1) to udbdat(1).
c userou is called only after all integration points
c are processed.
c 3. in post1, access udbdat(1) at the end of the nmisc record
c (miscellaneous nonsummable record).
```

```
c 4. multiply by the element volume and accumulate.
c this procedure presumes that the creep work density does not
c change rapidly in a region of irregular elements.
c
c this routine demonstrates the computation of creep strains using a
c primary creep equation equivalent to c6=0. see section 3.3 of the
ansys
c theoretical manual for more detailed documentation.
c in a large strain analysis, all strains are the hencky (i.e. log)
strains.
c all quantities are rotation-neutralized and are in the rotated
element
c coordinate system (not global).
c
c ***** initializations *****
c
c – initial checks
c – author should remove the warning below when making any
changes
if (intpt.eq.1 .and. kfirst.eq.1)
c
x call erhandler('usercr',5000,2,
x 'ANSYS,INC.-supplied version of coding for USERCR has been
x used.'
x ,0.0d0,' ')
c
c – check that author-defined usvr size is within bounds
c
if (nuval*nintp.gt.840)
Xcall erhandler('usercr',5010,4,
x 'Maximum storage allowed by USVR has been exceeded.'
x ,0.0d0,' ')
c
c – inititialize creep strain increment in case no creep
delcr = 0.0d0
c – no creep if time is not moving
if (timinc .le. 0.0d0 .and. timval .le. 0.0d0) go to 999
c
c ***** define the equivalent strain and stress *****
c
c – define the equivalent strain using the function egen
epet = egen (ncomp,epel(1),posn)
c – no creep if the strain is zero
if (epet .eq. 0.0d0) go to 999
c – define the stress
en=e*(1-0.44*(fre*timinc/nfinal))
sigen = en*epet
c
```

```
c ***** define the creep strain rate *****
para1=proptb(1)
smax=proptb(2)
fre=proptb(3)
nfinal=proptb(4)
para2=proptb(5)
c
c ***** normal beginning of user changes
c ***** beginning of sample creep law - same as c6 = 0
c
c — skip if c1 is zero
if (proptb(1) .eq. 0.0d0) go to 999
c
c — define the equivalent total strain
call vapb (epel(1),epcrp(1),ept(1),ncomp)
eptot = egen (ncomp,ept(1),posn)
c
c — define the creep strain rate
if (eptot .gt. 0.0d0) then
if(fre*timinc/nfinal<=1/10) then
delcr =
& para1*((1+20*(1.184/smax-1)-0.74037/smax)
& *(fre*timinc/nfinal)
& +(100*(1-(1.184/smax)+10*0.74037/smax)
& *(fre*timinc/nfinal)**2))-epet
Else if(1/10 <=fre*timinc/nfinal<= 8/10) then
delcr = para1*(1.184/smax+0.74037/smax*
& (fre*timinc/nfinal)-0.1)-epet
Else if(8/10 <=fre*timinc/nfinal<= 1) then
delcr =
& para1*((1.184/smax+0.74037/smax*
& (fre*timinc/nfinal)-0.1)+25*(
& (para2/para1)-1.184/smax-0.9*
& .74037/smax)*(fre*timinc/nfinal
& -0.8)**2)-epet
endif
endif
c
c ***** end of sample creep law
c ***** normal end of user changes
c
c — use prandtl-reuss relations to compute the increments for each
component
if (ncomp .eq. 1) then
del(1) = delcr*epel(1)/epet
else
con = delcr/epet
con = con/(2.0d0*(1.0d0+posn))
call vmult (epel(4),del(4),ncomp-3,3.0d0*con)
```

```
del(1) = con*(2.0d0*epel(1) - epel(2) - epel(3))
del(2) = con*(2.0d0*epel(2) - epel(3) - epel(1))
del(3) = con*(2.0d0*epel(3) - epel(1) - epel(2))
endif
c
c — update the strains
call vapb1 (epcrp(1),del(1),ncomp)
call vamb1 (epel(1),del(1),ncomp)
c
999 return
end
```

Appendix K: User subroutine for steel under cyclic load

This appendix provides the finite element code for the nonlinear analysis of steel under cyclic load, which is applied as a user-defined subroutine in ANSYS.

```
*deck,usermat3d USERDISTRIB parallel
gal
subroutine usermat3d(
& matId,elemId,kDomIntPt,kLayer,kSectPt,
& ldstep,isubst,keycut,
& nDirect,nShear,ncomp,nStatev,nProp,
& Time,dTime,Temp,dTemp,
& stress,ustatev,dsdePl,sedEl,sedPl,epseq,
& Strain,dStrain,epsPl,prop,coords,
& var0,defGrad_t,defGrad,
& tsstif,epsZZ,
& var1,var2,var3,var4,var5,
& var6,var7,var8)
c*****************************************************************
c *** primary function ***
c
c user defined material constitutive model
c
c Attention:
c User must define material constitutive law properly
c according to the stress state such as 3D, plane strain
c and axisymmetry, plane stress and beam.
c
c a 3D material constitutive model can use for
c plane strain and axisymmetry cases.
c
c When using shell elements, a plane stress algorithm
c must be use.
c
c gal July, 1999
c
c The following demonstrates a USERMAT subroutine for
c a plasticity model of 3D solid elements or plane elements
c in plane strain or axisymmetric stress state. The plasticity
```

```
c model is the same as TB, BISO.
c See "ANSYS user material subroutine USERMAT" for detailed
c description of how to write a USERMAT routine.
c
c*************************************************************
c
c input arguments
c ===============
c matId (int,sc,i) material #
c elemId (int,sc,i) element #
c kDomIntPt (int,sc,i) "k"th domain integration point
c kLayer (int,sc,i) "k"th layer
c kSectPt (int,sc,i) "k"th Section point
c ldstep (int,sc,i) load step number
c isubst (int,sc,i) substep number
c nDirect (int,sc,in) # of direct components
c nShear (int,sc,in) # of shear components
c ncomp (int,sc,in) nDirect + nShear
c nstatev (int,sc,l) Number of state variables
c nProp (int,sc,l) Number of material ocnstants
c
c Temp (dp,sc,in) temperature at beginning of time increment
c dTemp (dp,sc,in) temperature increment
c Time (dp,sc,in) time at beginning of increment (t)
c dTime (dp,sc,in) current time increment (dt)
c
c Strain (dp,ar(ncomp),i) Strain at beginning of time increment
c dStrain (dp,ar(ncomp),i) Strain increment
c prop (dp,ar(nprop),i) Material constants defined by TB,USER
c coords (dp,ar(3),i) current coordinates
c defGrad_t(dp,ar(3,3),i) Deformation gradient at time t
c defGrad (dp,ar(3,3),i) Deformation gradient at time t+dt
c
c input output arguments
c ======================
c stress (dp,ar(nTesn),io) stress
c ustatev (dp,ar(nstatev),io) user state variable
c ustatev(1) - equivalent plastic strain
c ustatev(2) - statev(1+ncomp) - plastic strain vector
c ustatev(nStatev) - von-Mises stress
c sedEl (dp,sc,io) elastic work
c sedPl (dp,sc,io) plastic work
c epseq (dp,sc,io) equivalent plastic strain
c tsstif (dp,ar(2),io) transverse shear stiffness
c tsstif(1) - Gxz
c tsstif(2) - Gyz
c tsstif(1) is also used to calculate hourglass
c stiffness, this value must be defined when low
c order element, such as 181, 182, 185 with uniform
```

```
c integration is used.
c var? (dp,sc,io) not used, they are reserved arguments
c for further development
c
c output arguments
c ================
c keycut (int,sc,io) loading bisect/cut control
c 0 - no bisect/cut
c 1 - bisect/cut
c (factor will be determined by ANSYS solution control)
c dsdePl (dp,ar(ncomp,ncomp),io) material jacobian matrix
c epsZZ (dp,sc,o) strain epsZZ for plane stress,
c define it when accounting for thickness change
c in shell and plane stress states
c
c*************************************************************
c
c ncomp 6 for 3D (nshear=3)
c ncomp 4 for plane strain or axisymmetric (nShear = 1)
c ncomp 3 for plane stress (nShear = 1)
c ncomp 3 for 3d beam (nShear = 2)
c ncomp 1 for 1D (nShear = 0)
c
c stresss and strains, plastic strain vectors
c 11, 22, 33, 12, 23, 13 for 3D
c 11, 22, 33, 12 for plane strain or axisymmetry
c 11, 22, 12 for plane stress
c 11, 13, 12 for 3d beam
c 11 for 1D
c
c material jacobian matrix
c 3D
c dsdePl | 1111 1122 1133 1112 1123 1113 |

c dsdePl | 2211 2222 2233 2212 2223 2213 |
c dsdePl | 3311 3322 3333 3312 3323 3313 |
c dsdePl | 1211 1222 1233 1212 1223 1213 |
c dsdePl | 2311 2322 2333 2312 2323 2313 |
c dsdePl | 1311 1322 1333 1312 1323 1313 |
c plane strain or axisymmetric (11, 22, 33, 12)
c dsdePl | 1111 1122 1133 1112 |
c dsdePl | 2211 2222 2233 2212 |
c dsdePl | 3311 3322 3333 3312 |
c dsdePl | 1211 1222 1233 1212 |
c plane stress (11, 22, 12)
c dsdePl | 1111 1122 1112 |
c dsdePl | 2211 2222 2212 |
c dsdePl | 1211 1222 1212 |
c 3d beam (11, 13, 12)
```

```
c dsdePl | 1111 1113 1112 |
c dsdePl | 1311 1313 1312 |
c dsdePl | 1211 1213 1212 |
c 1d
c dsdePl | 1111 |
c
c****************************************************************
#include "impcom.inc"
c
INTEGER
& matId, elemId,
& kDomIntPt, kLayer, kSectPt,
& ldstep, isubst, keycut,
& nDirect, nShear, ncomp, nStatev, nProp
DOUBLE PRECISION
& Time, dTime, Temp, dTemp,
& sedEl, sedPl, epseq, epsZZ
DOUBLE PRECISION
& stress (ncomp), ustatev (nStatev),
& dsdePl (ncomp,ncomp),
& Strain (ncomp), dStrain (ncomp),
& epsPl (ncomp), prop (nProp),
& coords (3),
& defGrad (3,3), defGrad_t(3,3),
& tsstif (2)
c
c**************** User defined part
*************************************
c
c - parameters
c
INTEGER mcomp
DOUBLE PRECISION HALF, THIRD, ONE, TWO, SMALL, ONEHALF,
& ZERO, TWOTHIRD, ONEDM02, ONEDM05, sqTiny, TWOFRE,
& BP, CP, AP, FP
PARAMETER (ZERO = 0.d0,
& HALF = 0.5d0,
& THIRD = 1.d0/3.d0,
& ONE = 1.d0,
& TWO = 2.d0,
& SMALL = 1.d-08,
& TWOFRE = 6.D0,
& BP = -0.09D0,
& CP = -0.56D0,
& AP = -0.53D0,
& FP = 0.832D0,
& sqTiny = 1.d-20,
& ONEDM02 = 1.d-02,
& ONEDM05 = 1.d-05,
```

```
& ONEHALF = 1.5d0,
& TWOTHIRD = 2.0d0/3.0d0,
& mcomp = 6
& )
c
c - local variables
c
c sigElp (dp,ar(6),l) trial stress
c dsdeEl (dp,ar(6,6),l) elastic moduli
c sigDev (dp,ar(6),l) deviatoric stress tensor
c dfds (dp,ar(6),l) derivative of the yield function
c JM (dp,ar(6,6),l) 2D matrix for a 4 order tensor
c pEl (dp,sc,l) hydrostatic pressure stress
c qEl (dp,sc,l) von-mises stress
c pleq_t (dp,sc,l) equivalent plastic strain at
beginnig of time increment
c pleq (dp,sc,l) equivalent plastic strain at
end of time increment
c dpleq (dp,sc,l) incremental equivalent plastic strain
c sigy_t (dp,sc,l) yield stress at beginnig of time increments
c sigy (dp,sc,l) yield stress at end of time increment
c young (dp,sc,l) Young's modulus
c posn (dp,sc,l) Poiss's ratio
c sigy0 (dp,sc,l) initial yield stress
c dsigdep (dp,sc,l) plastic slop
c twoG (dp,sc,l) two time of shear moduli
c threeG (dp,sc,l) three time of shear moduli
c
c - temperary variables for solution purpose
c i, j
c threeOv2qEl, oneOv3G, qElOv3G, con1, con2, fratio
c
EXTERNAL vzero, vmove, get_ElmData
DOUBLE PRECISION sigElp(mcomp), dsdeEl(mcomp,mcomp), G(mcomp),
& sigDev(mcomp), JM(mcomp,mcomp),
dfds(mcomp),
& sigi(mcomp), strainEl(mcomp)
DOUBLE PRECISION var0, var1, var2, var3, var4, var5,
& var6, var7, var8
DATA G/1.0D0,1.0D0,1.0D0,0.0D0,0.0D0,0.0D0/
c
INTEGER i, j
DOUBLE PRECISION pEl, qEl, pleq_t, sigy_t, sigy,
& dpleq, pleq,
& young, posn, sigy0, dsigdep,
& elast1, elast2,
& twoG, threeG, oneOv3G, qElOv3G, threeOv2qEl,
& fratio, con1, con2, dperr(3)
c***************************************************************
```

```
c
keycut = 0
dsigdep = ZERO
pleq_t = ustatev(1)
pleq = pleq_t
c *** get Young's modulus and Poisson's ratio, initial yield stress
and others
young = prop(1)
posn = prop(2)
sigy0 = prop(3)
c *** plastic strain tensor
call vmove(ustatev(2),epsPl(1),ncomp)
c *** calculate plastic slope
dsigdep = young*prop(4)/(young-prop(4))
twoG = young/(ONE+posn)
threeG = ONEHALF*twoG
elast1=young*posn/((1.0D0+posn)*(1.0D0-TWO*posn))
elast2=HALF*twoG
c *** define tsstif(1) since it is used for calculation of hourglass
stiffness
tsstif(1) = elast2
c
c *** calculate elastic stiffness matrix (3d)
c
dsdeEl(1,1)=(elast1+TWO*elast2)*G(1)*G(1)
dsdeEl(1,2)=elast1*G(1)*G(2)+elast2*TWO*G(4)*G(4)
dsdeEl(1,3)=elast1*G(1)*G(3)+elast2*TWO*G(5)*G(5)
dsdeEl(1,4)=elast1*G(1)*G(4)+elast2*TWO*G(1)*G(4)
dsdeEl(1,5)=elast1*G(1)*G(5)+elast2*TWO*G(1)*G(5)
dsdeEl(1,6)=elast1*G(1)*G(6)+elast2*TWO*G(4)*G(5)
dsdeEl(2,2)=(elast1+TWO*elast2)*G(2)*G(2)
dsdeEl(2,3)=elast1*G(2)*G(3)+elast2*TWO*G(6)*G(6)
dsdeEl(2,4)=elast1*G(2)*G(4)+elast2*TWO*G(1)*G(4)
dsdeEl(2,5)=elast1*G(2)*G(5)+elast2*TWO*G(1)*G(5)
dsdeEl(2,6)=elast1*G(2)*G(6)+elast2*TWO*G(2)*G(6)
dsdeEl(3,3)=(elast1+TWO*elast2)*G(3)*G(3)
dsdeEl(3,4)=elast1*G(3)*G(4)+elast2*TWO*G(5)*G(6)
dsdeEl(3,5)=elast1*G(3)*G(5)+elast2*TWO*G(5)*G(3)
dsdeEl(3,6)=elast1*G(3)*G(6)+elast2*TWO*G(6)*G(3)
dsdeEl(4,4)=elast1*G(4)*G(4)+elast2*(G(1)*G(2)+G(4)*G(4))
dsdeEl(4,5)=elast1*G(4)*G(5)+elast2*(G(1)*G(6)+G(5)*G(4))
dsdeEl(4,6)=elast1*G(4)*G(6)+elast2*(G(4)*G(6)+G(5)*G(2))
dsdeEl(5,5)=elast1*G(5)*G(5)+elast2*(G(1)*G(3)+G(5)*G(5))
dsdeEl(5,6)=elast1*G(5)*G(6)+elast2*(G(4)*G(3)+G(5)*G(6))
dsdeEl(6,6)=elast1*G(6)*G(6)+elast2*(G(2)*G(3)+G(6)*G(6))
do i=1,ncomp-1
do j=i+1,ncomp
dsdeEl(j,i)=dsdeEl(i,j)
end do
```

```
end do
c
c
c *** get initial stress
call vzero(sigi(1),ncomp)
i = ncomp
call get_ElmData ('ISIG',elemId,kDomIntPt,i,sigi)
c
c *** calculate the trial stress and
c copy elastic moduli dsdeEl to material Jacobian matrix
do i=1,ncomp
strainEl(i) =
& Strain(i)+0.623*(dStrain(i))**FP
& *(TWOFRE*(Time+ONEDM02))**BP
& -0.0196*(epspl(i))**(0.155)*(dStrain(i))**AP
& *(TWOFRE*(Time+ONEDM02))**CP
end do
call vzero(sigElp,6)
do i=1,ncomp
do j=1,ncomp
dsdePl(j,i) = dsdeEl(j,i)
sigElp(i) = sigElp(i)+dsdeEl(j,i)*strainEl(j)
end do
sigElp(i) = sigElp(i)+sigi(i)
end do
c *** hydrostatic pressure stress
pEl = -THIRD*(sigElp(1)+sigElp(2)+sigElp(3))
c *** compute the deviatoric stress tensor
sigDev(1) = sigElp(1)+pEl
sigDev(2) = sigElp(2)+pEl
sigDev(3) = sigElp(3)+pEl
sigDev(4) = sigElp(4)
sigDev(5) = sigElp(5)
sigDev(6) = sigElp(6)
c *** compute von-mises stress
qEl =
& sigDev(1)*sigDev(1)+sigDev(2)*sigDev(2)+
& sigDev(3)*sigDev(3)+
& TWO*(sigDev(4)*sigDev(4)+sigDev(5)*sigDev(5)+
& sigDev(6)*sigDev(6))
qEl = sqrt(ONEHALF*qEl)
c *** compute current yield stress
sigy = sigy0+dsigdep*pleq
c
fratio = qEl/sigy-ONE
c *** check for yielding
IF (sigy .LE. ZERO.or.fratio .LE. -SMALL) GO TO 500
c
sigy_t = sigy
```

```
threeOv2qEl = ONEHALF/qEl
c *** compute derivative of the yield function
DO i=1, ncomp
dfds(i) = threeOv2qEl*sigDev(i)
END DO
oneOv3G = ONE/threeG
qElOv3G = qEl*oneOv3G
c *** initial guess of incremental equivalent plastic strain
dpleq = qElOv3G-sigy*oneOv3G
pleq = pleq_t+dpleq
sigy = sigy0+dsigdep*pleq
c
c *** update stresses
DO i = 1, ncomp
stress(i) = sigElp(i)-TWOTHIRD*(qEl-sigy)*dfds(i)
END DO
c
c *** update plastic strains
DO i = 1, nDirect
epsPl(i) = epsPl(i)+dfds(i)*dpleq
END DO
DO i = nDirect+1, ncomp
epsPl(i) = epsPl(i)+TWO*dfds(i)*dpleq
END DO
epseq = pleq
c *** Update state variables
ustatev(1) = pleq
do i=1,ncomp
ustatev(i+1) = epsPl(i)
end do
c *** Update plastic work
sedPl = sedPl+HALF*(sigy_t+sigy)*dpleq
c
c *** Material Jcobian matrix
c
IF (qEl.LT.sqTiny) THEN
con1 = ZERO
ELSE
con1 = threeG*dpleq/qEl
END IF
con2 = threeG/(threeG+dsigdep)-con1
con2 = TWOTHIRD*con2
DO i=1,ncomp
DO j=1,ncomp
JM(j,i) = ZERO
END DO
END DO
DO i=1,nDirect
DO j=1,nDirect
```

```
JM(i,j) = -THIRD
END DO
JM(i,i) = JM(i,i)+ONE
END DO
DO i=nDirect+1,ncomp
JM(i,i) = HALF
END DO
DO i=1,ncomp
DO j=1,ncomp
dsdePl(i,j) = dsdeEl(i,j)-twoG
& *(con2*dfds(i)*dfds(j)+con1*JM(i,j))
END DO
END DO
c
goto 600
500 continue
c *** Update stress in case of elastic/unloading
do i=1,ncomp
stress(i) = sigElp(i)
end do
600 continue
sedEl = ZERO
DO i = 1,ncomp
sedEl = sedEl+stress(i)*(Strain(i)+dStrain(i)-epsPl(i))
END DO
sedEl = sedEl*HALF
ustatev(nStatev) = sigy
c
return
end
```

Appendix L: User subroutine for FRP under cyclic load

This appendix provides the finite element code for analysis of FRP under cyclic load, which is applied as a user-defined subroutine in ANSYS.

```
*deck,usermat USERDISTRIB parallel
gal
subroutine usermat(
& matId,elemId,kDomIntPt,kLayer,kSectPt,
& ldstep,isubst,keycut,
& nDirect,nShear,ncomp,nStatev,nProp,
& Time,dTime,Temp,dTemp,
& stress,ustatev,dsdePl,sedEl,sedPl,epseq,
& Strain,dStrain,epsPl,prop,coords,
& var0,defGrad_t,defGrad,
& tsstif,epsZZ,
& var1,var2,var3,var4,var5,
& var6,var7,var8)
c*************************************************************
#include "impcom.inc"
c
INTEGER
& matId, elemId,
& kDomIntPt, kLayer, kSectPt,
& ldstep, isubst, keycut,
& nDirect, nShear, ncomp, nStatev, nProp
DOUBLE PRECISION
& Time, dTime, Temp, dTemp,
& sedEl,sedPl, epseq, epsZZ
DOUBLE PRECISION
& stress(ncomp), ustatev(nStatev),
& dsdePl(ncomp,ncomp),
& Strain(ncomp), dStrain(ncomp),
& epsPl(ncomp), prop(nProp),
& coords(3),
& defGrad(3,3), defGrad_t(3,3),
& tsstif(2)
```

```
C
DOUBLE PRECISION var0, var1, var2, var3, var4, var5,
& var6, var7, var8
C*********************************************************
INTEGER K1, K2
DOUBLE PRECISION ONE, TWO, FRE
PARAMETER (ONE=1.0D0, TWO=2.0D0, FRE=3.0D0)
DOUBLE PRECISION EMOD, ENU, EG, EG2, ELAM, EC
DOUBLE PRECISION SesSrn(6,6)
EMOD = prop(1)
ENU = prop(2)
EC = EMOD-1100*log((Time+1)*FRE)
EG = EC/(TWO*(ONE+ENU))
EG2 = TWO*EG
ELAM = EC*ENU/(ONE+ENU)/(ONE-TWO*ENU)
C the elastic modulus is changed with the time varies
C the time must be greater than zero otherwise the EG is going infinity
C
C
DO K1=1,3
DO K2=1,3
SesSrn(K2,K1)=ELAM
END DO
SesSrn(K1,K1)=EG2+ELAM
END DO
DO K1=4,6
SesSrn(K1,K1)=EG
END DO
DO K1=1,6
DO K2=1,6
stress(K2)=stress(K2)+SesSrn(K2,K1)*dStrain(K1)
END DO
END DO
DO K1=1,6
DO K2=1,6
dsdePl(K1,K2)=SesSrn(K1,K2)
END DO
END DO
END IF
return
end
```

Index

Note: Page numbers followed by *f* indicate figures and *t* indicate tables.

Kolibri 360 GmbH
Ferdinand-Jühlke-Straße 7
99095 Erfurt, Deutschland
produktsicherheit@kolibri360.de